Eco-Crisis

Eco-Crisis

Cecil E. Johnson
Riverside City College
Riverside, California

John Wiley & Sons, Inc., New York . London . Sydney . Toronto

Library of Congress Catalogue Card Number: 79-136761

ISBN 0 471 44296 8

Printed in the United States of America

10 9 8 7 6 5 4 3 2 1

Preface

This book of readings deals with the ecological realities that today's college students will have to face. The immediate danger is that the human race will wipe itself out with its H-bombs or its methods of chemical and biological warfare. This risk is generally recognized.

But most people tend to ignore another threat, almost as grave, which is posed by the fact that the population on our planet will double by the year 2000. This means that most of humanity, except for a well-fed minority of the powerful, will endure a starving and miserable existence. Nothing is more likely to lead to confrontation between the major powers than the threat of universal blight through over population.

These selections also examine the poblems of dirty air, polluted water, and the gradual encroachment of litter across the land. A chance to enjoy inner peace in the world of nature is hard to come by, especially near the large metropolitan centers; and some of our last wilderness areas are threatened by bulldozers. The public heritage may, if we do not defend it, fall prey to privated greed.

Finally, our very health and sanity are undermined by the noise and stress of modern life. The time has come to do something about it.

This book is not designed for the weak of heart who continually see their world through rose-colored glasses. By grasping reality as it is encompassed in these selections, we may learn how to face these problems and resolve them. If we do not, there is little hope for the future. All these ecological problems press in on us, impoverish the human spirit, and may, in the end, destroy the human environment.

Cecil E. Johnson
Riverside, California

Contents

Contents

Eco-Crisis

Introduction

Recently I was sitting in the high mountain country not far from Los Angeles. The sky above was a beautiful blue, and the orange-red sun seemed occasionally to caress the fluffy snow-white thunderheads. A summer rainstorm had just passed over and the dead, sun-browned pine needles, which matted the ground, soaked up the water like hungry sponges. The western yellow pines, incense cedars, canyon live oaks, and white firs all sent out their spicy odors as they cast mottled patches of shade on the earth below.

I was momentarily consumed by the sights, smells, and utter grandeur of it all. Reluctantly, I turned my eyes to the clouds overhanging the basin below, in which numerous look-alike cities teemed with millions of human beings. Each city appeared trapped in a gargantuan plastic bag of suffocating smog.

The razor-sharp contrast between the pure air and blue sky around my mountain camp and the ominous, dark clouds far below filled me with a sense of impending disaster. It seemed paradoxical that *Homo sapiens* ("wise man") had just landed two men on the moon but would not deal with the environmental garbage dump that sprawled, spread, and scattered in all directions below.

Slowly I took down my tent and stored it in my car with other paraphernalia. Then, firing up my internal-combustion engine, I headed back for my smog-ridden suburbia far below. My car, even though made of metal, leather, rubber, and plastic doodads, seemed almost alive as she belched out her share of pollutants. Then, together, we slowly began to weave our way back home.

I noticed with a simmering resentment the shiny, discarded beer cans and the heat-reflecting corpses of soft-drink bottles along the roadside. The weekend picnickers had just visited the area the day before. Their discarded paper napkins, cellophane sandwich bags, and gaudily colored potato-chip containers rolled down the road like tumbleweeds. Pieces of cast-off Styrene cups dot the roadside. They also lie in endless battalions along our seashores, rivers, and lakes. These cups are cheap to manufacture, easy to throw away, and difficult to break down by nature's methods. Since it is unlikely that the American public's littering habits are going to change, a different approach must be taken. An attack must be mounted against the technological establishments that have made these eyesores possible. Professor Samuel F. Huelbert of Clemson University is currently on the research trail of a bottle that, when broken, turns soft in a few days and then dissolves into a pool of water. Yearly, the American public throws away about 26 billion bottles. Even more cans are discarded. Their ugly carcasses not only blight the landscape but also represent a tremendous waste of natural resources.

At about 1500 feet about sea level, my car punctured its way back into the smog zone. I then entered the web of freeways that leads to the City of Angels, Los Angeles. My eyes began to water and my lungs cried out for the clean mountain air still perched precariously 5000 feet above me. It was 4:30 P.M. and numerous species of cars were packed bumper to bumper. As I crawled along, I saw those perennial enemies of nature, the bulldozers. Like giant metal beetles, they were doing their bit to extend the monstrous freeway system which enmeshes not only Los Angeles but all our major cities.

Shopping centers, surrounded by vast, barren stretches of asphalt for parking, litter the landscape. They get their color from modern man's most ubiquitous "work of art"—the highway billboard. Gasoline stations, laundromats, restaurants, bars, supermarkets, and innumerable other structures help complete this picture of roadside beauty. During the vacation season, tension-ridden city dwellers and suburbanites join the annual mass exodus from their plastic, steel, and concrete world. Often they hope to camp alone with nature in some beautiful area; but they usually end up crowded together back to back like sardines in a can. Instead of enjoying peaceful evenings by a campfire, they will, more than likely, hear blaring transistor radios and howling motorcycles pounding incessant commercial rhythms into their skulls.

Some state and national park visitors want to escape this backyard-barbecue atmosphere. They hardy people, sometimes whole families, back-pack into our wilderness areas, which offer them the challenge of physical exertion as well as the sounds, smells, sights, and feel of nature in the raw. Dr. Howard Zahniser's contribution, reflecting his acute awareness of the relatively small amount of wilderness area left, is a classic of its kind: "A wilderness, in contrast with those areas where man and his own works dominate the landscape, is hereby recognized as an area where man himself is a visitor who does not remain."

Professor Garrett Hardin's eloquent, realistic, hard-hitting essay, entitled "The Economics of Wilderness" makes a great deal of sense. He says "There is .. a public interest in making the wilderness as difficult and dangerous as it legitimately can be. There is, I think, a well-founded suspicion that our life has become, if anything, too safe for the best psychological health, particularly among the young." He comments further, "A wilderness that can be entered only by a few of the most physically fit of the population will act as an incentive to myriads more to improve their physical condition." "This does not mean," Professor Hardin continues, "[that] we cannot expand our present practice of setting up small outdoor areas where we permit a high density of people to get a tiny whiff of nature .. there are people who simply love this slumming togetherness."

The eminent humanist-biologist, Paul Ehrlich, in the *The Population Bomb,* his latest book, has placed his sensitive scientific fingertips on the whole human ecology in what he calls "Mankind's Inalienable Rights:"

1. The right to eat well.
2. The right to drink pure water.
3. The right to decent, uncrowded shelter.
4. The right to enjoy natural beauty.
5. The right to avoid pesticide poisoning.
6. The right to freedom from thermonuclear war.
7. The right to limit families.

I would like to include another right:

8. The right to peace and quiet.

The preservation of environmental quality and the fulfillment of man's desire to live in a world of natural beauty will require continual vigilance and the coordinated efforts of the federal government, individual states, industry, and the general public. If we fail to develop an ecological approach in planning the future, our descendants may live in a drab, colorless tomorrow, a world of endless blight. Currently we have a remarkable ecological grass-roots movement surging like a giant undercurrent throughout the United States. If the momentum can be maintained, the country may yet keep from losing the race to preserve surroundings in which man can live. May the ecologists always be in a position, unhampered by the vested interests of commercialism, to serve science and mankind. We must first open our eyes and see the harsh reality; but we must also understand nature's delicate balance. Man cannot continue to perpetrate irreversible acts of destruction against our all-giving mother, the earth, and against the sky that surrounds her.

1.

Can the World Be Saved?

LaMont C. Cole

Professor LaMont C. Cole is rare among men of genius because he has a gift for communicating his ideas to the lay public.

He is one of America's most outstanding scientists and is currently Professor of Ecology and Systematics, Division of Biological Sciences, at Cornell University.

This essay encompasses discussions of the impact of man on the environment, the "backward countries," the hazards of environmental deterioration, our multiple cultures, and population regulation and prospects for the future.

My title here is not my first choice, but a year or so ago a physicist discussing some of the same subjects (Berkner, 1966) beat me to the use of the title I would have preferred: "Is There Intelligent Life on Earth?" However, the present title may have merit if it attracts just a few persons who are looking for something about either religion or warfare—because such persons constitute an audience I would like to reach.

In recent years we have heard much discussion of distinct and nearly independent cultures within our society that fail to communicate with each other. This strikes me forcibly, for example, almost every time I hear natural scientists attempting to communicate with social scientists, and we surely have many more than two cultures with such difficulties of communication.

The dichotomy I am concerned with here is that between ecologists and their fellow thinkers on the one hand and, on the other, persons who consider continuous growth desirable—growth of population, of government, industry, trade, and even agriculture. This is also, in part, a dichotomy between the thinkers and the doers—those who insist that man should try to know the consequences of his actions before he takes them versus those who want to get on with the building of dams and canals, with the straightening of river channels, with the firing of nuclear explosives, and with the industrialization of backward countries simply because they know how to do these things.

The Background

To develop my theme, I shall have to start far back in time. When the world was young, it did not have an atmosphere of the gases that are now present. The water that now fills the oceans and furnishes our precipitation and the nitrogen that makes up most of the present atmosphere originated from rocks in the earth's interior by a degassing process (Rubey, 1955).

The amount of oxygen in the atmosphere was negligible before the origin of living organisms that could carry on photosynthesis of the type characterizing green plants (Urey, 1959, Berkner and Marshall, 1965), and oxygen would disappear from the atmosphere now

Reprinted by permission of the author, LaMont C. Cole, and the publisher, BioScience. Copyright American Institute of Biological Sciences. Paper presented at the 134th meeting of the American Association for the Advancement of Science, December 27, 1967.

through natural geological processes if all the green plants should be killed (Clarke, 1920).[1] In the early days of photosynthesis, oxygen was used as rapidly as it was produced, so there was virtually no accumulation of a reservoir in the atmosphere. But, very gradually, some of the compounds produced by organisms began to pass out of circulation by being deposited in sedimentary rocks in such forms as crude oil. With these compounds protected from oxidation, a small percentage of oxygen became a regular constituent of the atmosphere. Eventually, perhaps not until Carboniferous times only three or four hundred million years ago, the deposition of unoxidized organic matter, partly in the form of what we call the fossil fuels, coal, petroleum, and natural gas brought the oxygen content of the atmosphere to approximately its present level of slightly over 20%.

The combination of green plants and oxidizing organisms, including animals, apparently became very efficient at taking oxygen from the atmosphere and returning it at equal rates so that the amount present has remained virtually constant. Photosynthesis stops, of course, during the hours of darkness, and on land areas in high latitudes it practically stops during the winter. It continues, however, in low latitudes (although often greatly reduced by seasonal drought) and in the oceans and, fortunately, atmospheric circulation patterns are such that we have not yet had to be concerned that man would run out of oxygen to breathe at night or in winter.

The atmosphere and the bodies of organisms are the earth's only important reservoirs of the nitrogen which all organisms require for building proteins, and in this case also the totality of life was able to balance the rates of use and return so that the percentage of nitrogen in the atmosphere remains constant. This, however, is a much more complex process than the maintenance of oxygen.

[1]Several investigators have concluded that the atmosphere is very slowly losing its oxygen to these processes despite photosynthesis.

Certain bacteria and primitive algae absorb nitrogen and convert it to ammonia. Others convert the ammonia to nitrate. Green plants use both ammonia and nitrate to build their proteins, and animals and microorganisms build all of their proteins from the constituents of plant proteins. Then a variety of decay organisms degrade the nitrogen compounds of dead plants and animals to simpler compounds, and additional kinds of microorganisms regenerate the molecular nitrogen in the atmosphere. If any one of the numerous steps in this nitrogen cycle were to be disrupted, disaster would ensue for life on earth. Depending upon which step broke down, the nitrogen in the atmosphere might disappear, it might be replaced by poisonous ammonia, or it might remain in the atmosphere with life disappearing for want of a way to use it in building proteins.

There are many other chemical elements in addition to oxygen and nitrogen which are required by organisms. Some of these also undergo fascinating cycles and a few, notably phosphorous and potassium, are gradually washing into the oceans to remain there in sediments until major geological upheavals, or possibly human ingenuity, can retrieve them and so perpetuate life on earth.

Here I shall mention only one additional cyclic chemical element, carbon, which is the distincitve constituent of all organic compounds. Its great reservoir is in the oceans which dissolve carbon dioxide from the atmosphere and precipitate it as limestone. This, however, is a slow process depending, in part, on the rate of turnover of deep ocean water (Bolin and Eriksson, 1959). Plants use carbon dioxide to build their organic compounds and animals combine the organic compounds with oxygen to obtain the energy for their activities. And this is possible only because, millions of years ago, the deposition of organic matter in sedimentary rocks created a reservoir of oxygen in the atmosphere.

The Impact of Man

Man existed on earth before the Pleistocene ice ages, but his numbers then were certainly negligible. Primitive hunters and food gatherers probably used fire to drive game and in so doing altered much of the earth's vegetation. Many of our major grassland areas have climates that would permit the growth of forests were it not for recurring fires and, more recently, heavy grazing and cultivation by man. These earlier grasslands included much of what has become the world's best agricultural land; therefore these early hunting cultures set the stage for the next phase in man's development—pastoral life and agriculture.

Early agricultural man concentrated his efforts on the flood plains of rivers where the soil was fertile and well watered and easy to cultivate with simple tools. It was discovered that certain types of food could be stored so that the produce of the growing season could support man and his domestic animals throughout the year. Then man started to build towns and cities and to expand his numbers. Crowding predisposed man to epidemics and these, together with wars and famines, took a frightful toll of human life. Nevertheless, the total human population continued to increase and to discover and exploit new lands.

I am referring here to times many centuries before the industrial revolution and the population explosion which are the forces behind our present world crisis. Plato knew that deforestation and overgrazing could cause soil erosion that would ruin fertile land and, at least a millennium before his time, great civilizations had destroyed themselves by means of what turned out to be unsound agricultural practices. But our present troubles stem from the time when man began serious exploitation of the fossil fuels. First it was coal that was important and now it is petroleum, but inexorably man is extracting the fossil fuels and recombining their carbon with oxygen. And this exploitive way of life permits, for the moment, so many more people to exist on earth simultaneously than has ever been possible before that man seems unwilling to consider what he is really doing to the earth.

"Backward Countries"

We hear a lot today about "underdeveloped" and "developing" nations, but these actually tend to be overdeveloped nations (Borgstrom, 1965).

The valleys of the Tigris and Euphrates supported the Sumerian civilization in 3500 B.C. A great irrigation complex was based on these rivers by 2000 B.C., and this was the granary of the great Babylonian Empire. Pliny tells of their harvesting two annual crops of grain and grazing sheep on the land between crops. But less than 20% of the land in modern Iraq is cultivated and more than half of the national income is from oil. The landscape is dotted with mounds representing forgotten towns, the ancient irrigation works are filled with silt, the end product of soil erosion, and the ancient seaport of Ur is now 150 miles from the sea with its buildings buried under as much as 35 feet of silt.

Similar conditions prevail in Iran which was once the seat of the great Persian empire and where Darius I was the "King of Kings" 2400 years ago. The present Shah is making a determined effort to rejuvenate this land, and this may be the test case of what the industrial world can do in repairing the ravages of the past.

The valley of the Nile was another cradle of civilization. Every year the river overflowed its banks at a predictable time, bringing water to the land and depositing a layer of silt rich in mineral nutrients for plants. Crops could be grown for 7 months each year. Extensive irrigation systems were established before 2000 B.C. This land was the granary of the Roman Empire, and this type of agriculture flourished for another 2000 years. But the population has continued to grow and economic considerations have diverted land from growing food to cash crops such as cotton. Then, in 1902 a dam was built at Aswan to prevent the spring flood and permit year-round irrigation. Since then the soils have been deteriorating through salinization, and productivity has decreased. The new Aswan high dam is designed to bring another million acres of land under irrigation, and it may well prove to be the ultimate disaster for Egypt. Meanwhile, population growth has virtually destroyed any possibility that the new agricultural land can significantly raise the average level of nutrition.

Sorry stories like this could be told for country after country. The glories of ancient Mali and Ghana in west Africa were legends in medieval Europe. Ancient Greece had forested hills, ample water, and productive soils. In the land that once exported the cedars of Lebanon to Egypt, the old Roman roads which have prevented erosion of the soil beneath them now stand several feet above the rock desert. But in a church yard that had been protected from goats for 300 years, the cedars were found about 1940 to be flourishing as in ancient times (Lowdermilk, 1948). Where there is soil left, this country could evidently be rehabilitated. In China and India, ancient irrigation systems stand abandoned and filled with silt. When the British assumed the rule of India two centuries ago, the population was about 60 million. Today, it is about 500 million and most of its land problems have been created in the past century through deforestation and plowing and the resulting erosion and siltation, all stemming from efforts to support this fantastic population growth.

In southern and eastern Asia there is a tendency to blame destruction of the land on invaders, in particular, Genghis Khan in the 13th Century and Tamerlane in the 14th. But engineering works can be rebuilt if it is worth doing so. The Romans cultivated mountain slopes and thus caused erosion that clogged streams and produced marshes which, as sources of malaria, are thought by some to have had more than a little to do with the decline of the Empire. But these lands have been reclaimed, as have comparable areas in France where Napoleon's engineers succeeded in reclaiming land from swamps and sand dunes which were created when the Vandals burned forests in the 5th Century.

Overdevelopment by man is not confined to the classical world. Gambia, with 96% of its income from peanuts, has very little left to develop. Archaeologists have long wondered how the Mayas managed to support what was obviously a high civilization on the now unproductive soils of Guatemala and Yucatan. Evidently, they exploited their land as intensively as possible until both its fertility and their civilization collapsed. In parts of Mexico the water table has fallen so that towns originally located to take advantage of superior springs now must carry in water from distant sites. As recently as the present decade, aerial reconnaissance has revealed ancient ridged fields on flood plains, the remnants of "a specialized system of agriculture that physically reshaped large parts of the South American continent" (Parsons and Denevan, 1967), and, even more recently, the same system of constructing ridges on seasonal swamps to permit agriculture free from either waterlogged soils or seasonal drought has been described from Tanzania in Africa (Deshler, 1967). The South American ridges occur in areas that have been considered unfit for agriculture and, indeed, salinization, the ancient curse of irrigation systems, has ruined the potential of at least some of these areas. In Africa, this practice of mounding or ridging soil to grow root crops is known to accelerate erosion (Nye and Greenland, 1960).

Even our own young country is not immune to deterioration. We have lost many thousands of acres to erosion and gullying, and many thousands more to strip mining. It has been estimated that the agricultural value of Iowa farmland, which is about as good land as we have, is declining by 1% per year. In our irrigated lands of the West, there is the constant danger of salinization from rising water tables while, elsewhere, from Long Island to Southern California, we have lowered water tables so greatly that in coastal regions salt water is seeping into the aquifers. Meanwhile, an estimated 2000 irrigation dams in the United States are now useless impoundments of silt, sand, and gravel.

So this is the heritage of man's past. Do we know better today? Two recent publications sponsored by the Federal government (Pres. Sci. Advis. Com., 1965; Linton, 1967) have examined in detail our damage to our environment, especially by pollution. I do not wish to cover the same ground here except to the extent that some current practices and proposals are relevant to the question of whether or not we can mend our ways while, and if, there is still time to save our civilization. I shall not deal here with direct hazards of pollution to man and his domesticated plants and animals. These are very serious problems but less alarming, in my opinion, than some of the larger and less obvious threats.

Hazards of Environmental Deterioration

Every year we are destroying fossil fuels at a greater rate than in the preceding year while, in this country alone, we are annually removing, largely by paving, a million acres from the cycle of photosynthetic productivity. We do not know to what extent we are inhibiting photosynthesis in either freshwater or marine environments. Thus, while we are accelerating the recombination of fossil carbon with oxygen, we are reducing the rate at which the oxygen in the atmosphere is regenerated.

When, and if, we reach the point where the rate of combustion exceeds the rate of photosynthesis, we shall not only have to worry about running out of oxygen at night and in winter, but the oxygen content of the atmosphere will actually decrease. If this occurred gradually, its effect would be approximately the same as moving everyone to higher altitudes, a change that might help to alleviate the population crisis by raising death rates. I am told, however, that the late Lloyd Berkner (1966) was less optimistic and thought that atmospheric depletion might occur suddenly and disastrously.

It is true that 70% or more of the total oxygen production by photosynthesis occurs in the oceans and is largely produced by planktonic diatoms. It is also true that we are dumping into the oceans vast quantities of pollutants consisting, according to one estimate by the U.S. Food and Drug Administration (Linton, 1967), of as many as a half-million substances. Many of these are of recent origin and are biologically active materials such as pesticides, radioisotopes, and detergents to which the earth's living forms have never before had to try to adapt. No more than a minute fraction of these substances and combinations of them has been tested for toxicity to marine diatoms or, for that matter, to the equally vital forms involved in the cycles of nitrogen and other essential elements. I do not think we are in a position to assert right now that we are not poisoning the marine diatoms and thus bringing disaster upon ourselves. If the tanker *Torrey Canyon* had been carrying a concentrated herbicide instead of petroleum, could photosynthesis in the North Sea have been stopped? Berkner is said to have considered that a very few such disasters occurring close enough together in time might cause the ultimate disaster. We must have our green plants both on land and in the sea, and I am happy that, so far, schemes such as a UNESCO plan of 20 years ago to "develop" the Amazon basin have been judged impracticable. Surely man's influence on earth is now so predominant that he must stop trusting to luck that he will not upset any of the indispensable biogeochemical cycles.

Another aspect of the combustion of fossil fuel is that we are adding carbon dioxide to the atmosphere more rapidly than the oceans can assimilate it (Bolen and Eriksson, 1959). Both carbon dioxide and water vapor in the atmosphere are more transparent to shortwave solar radiation than to the longwave heat radiation from the earth to space. They tend to raise the surface temperature and they have the potentiality for altering the earth's climates in ways that are still highly controversial, but which, everyone agrees, are undesirable.

Industrial plants, automobiles, and private homes are the big consumers of fossil fuels. To appreciate the magnitude of the problem, however, let us consider very briefly a still minor source of atmospheric pollution, the airplane. It may have disproportionate importance because much of its carbon dioxide and water are released at high altitude where they can take a long time to be removed from the atmosphere. When you burn a ton of hydrocarbon, you obtain as by-products about one and one-third tons of water and about twice this amount of carbon dioxide. A Boeing 707 in flight does this roughly every 10 minutes. I read in the papers that 10,000 airplanes per week land in New York City alone, and this does not include military planes. If we assume very crudely that the 707 is typical of airliners and that its average flight takes 4 hours, this amounts to an annual release into the atmosphere of about 18 million tons of water and twice that amount of carbon dioxide. And not all flights terminate in New York! How long can this go on or how much can it be expanded before we seriously alter the radiation balance of the earth and atmosphere?

In any case, if we do not destroy ourselves first, we are going to run out of fossil fuels, and this prospect is surely not many generations away. Then, presumably, we shall have to turn to atomic energy and face agonizing problems of environmental pollution. One would think that the present custodians of this vital resource for the future, which is also a nonrenewable resource, could think of better things to do with it than creating explosions.

I am aware that reactors to produce electricity are in use and under development, but I am apprehensive of what I know of the present generation of reactors and of those proposed for the future. The fuel for present reactors has to be reprocessed periodically. This yields long-lived and biologically hazardous isotopes such as ^{90}Strontium and ^{137}Cesium that should be stored where they cannot contaminate the environment for at least 1000 years; but a fair proportion of the storage tanks employed so far are leaking after only about 20 years (Snow, 1967; Belter and Pearce, 1965). This process also releases

^{85}Krypton into the atmosphere to add to the radiation exposure of the earth's biota including man, and I do not think that anyone knows a practicable way to prevent this. We are glibly offered the prospect of "clean" bombs and thermonuclear power plants which would not produce these isotopes, but, to the best of my knowledge, no one yet knows how this is to be accomplished. And, if development is successful, these reactors will produce new contaminants, among others tritium (^{3}Hydrogen) which becomes a constituent of water, in this case long-lived radioactive water, which will contaminate all environments and living things (Parker, 1968).[2] Even in an official publication of the Atomic Energy Commission it is suggested that for certain mining operations it may be better to use fission (i.e., "dirty") devices instead of fusion (i.e., "clean") devices "to avoid ground water contamination or ventilation problems" (Frank, 1964).

As a final example of what we may irresponsibly do to the world environment, let me mention the proposed new sea-level canal across Central America. In that latitude the Pacific Ocean stands higher than the Atlantic by a disputed amount which I believe to average 6 feet. The tides are out of phase on the two sides of the Isthmus of Panama so the maximum difference in level can be as great as 18 feet. Also, the Pacific is much colder than the Atlantic as a result of current patterns and the upwelling of cold water. If the new canal should move a mass of very cold water into the Caribbean, what might this do to climates, or to sea food industries? Nobody has the information to give an authoritative answer, but I have heard suggestions of a new hurricane center, or even diversion of the Gulf Stream with a resultant drastic effect on the climates of all regions bordering the North Atlantic.

We know that the sea-level Suez Canal permitted the exchange of many species between the Red Sea and the Mediterranean, and we know that the Welland Canal let sea lampreys and alewives enter the upper Great Lakes with disastrous effects on fisheries and, more recently, on bathing beaches. What will a tropical sea-level connection between the Atlantic and Pacific cause? Nobody knows.

Even more alarming to me is the prospect that nuclear explosives might be used to dig the canal because, in terms of immediate costs, this is evidently the most economical way. If 170 megatons of nuclear charges will do the job, as has been estimated by the Corps of Engineers which apparently wants to do the job (Graves, 1964), we can again take figures from an official Atomic Energy Commission publication (Stead, 1964) and see what this means in terms of environmental contamination with radioisotopes, assuming the explosions to take place in average materials of the earth's crust. For ^{137}Cesium alone there would be produced 2.72×10^{5} curies and, since the permissible whole-body human exposure to this isotope is set at 3×10^{6} curies, this amounts to 26.5 limiting doses for every one of the 3.4×10^{9} persons on earth. Cesium behaves as a gas in a cratering explosion and prevailing winds in the region are from East to West, so the Pacific would presumably be contaminated first. And cesium behaves like potassium in biological food chains, so we could anticipate its rapid dissemination among living things.

[2]Since this was originally written, it has been calculated that the amount of tritium released from fusion reactors, if these were to replace our other electrical generators, "would result in unacceptable worldwide dosages by the year 2000." (See Parker, 1968.) Even more recently this statement has been hedged and confidence expressed that a solution can be found. (See Parker and Rose, 1968.)

Our Multiple Cultures

This brings me back to our multiple subcultures and their failure of intercommunication. I do not want to comment on the advertising executive who asserts that billboards are "the art gallery of the public" or the spokesman of industry who said that "the ability of a river to absorb sewage is one of our great natural resources and should be utilized to the utmost." And I feel a little sorry for many of my friends who try to promote conservation on esthetic grounds, because I have come to suspect that those of us who care about open spaces and natural beauty are an incomprehensible minority. Are we selecting for genotypes those who can satisfy their esthetic needs in big city slums? Are the Davy Crocketts and Kit Carsons who are born today destined for asylums, jail, or suicide? I am afraid this depressing thought may contain an element of truth. However, it is clear to me that the fundamental problem in the world today is simply too many people multiplying too rapidly and placing demands on the world's resources that cannot be sustained.

It alarms me to read almost daily suggestions as to how food production can be increased "to keep up with the growing population," but only infrequently, and usually in obscure places, to come across authors who recognize the obvious fact that it is impossible to provide food for a population that continues to grow exponentially as ours is now doing at a rate of 1.7% per year. Recently, I have been hearing suggestions for using bacteria, fungi, or yeasts to convert petroleum directly into food for man. This is superficially attractive because it appears to be more efficient than first feeding the petroleum to a refinery and then to tractors and other machines which eventually deliver food to us. It is a melancholy fact that the metabolism of bacteria, fungi, and yeasts does not generate oxygen.

Population Regulation

There appears to be no way for us to escape our dependence on green plants and no way for us to survive except to halt population growth completely or even to undergo a period of population decrease if, as I anticipate, definitive studies show our population to be already beyond what the earth can support on a continuous basis. In order to accomplish this, natural scientists, social scientists, and political leaders will have to learn to communicate with each other and with the general public. This is a large order, but I have learned in recent years that intercommunication is possible with social scientists who are concerned with population problems, and I am now hopeful that these subcultures can come to understand each others' problems and to appreciate efforts to solve them. Even some of the extreme optimists, who believe that it would be possible for some time to make world food production increase more rapidly than the human population, agree that population limitation is necessary to prevent other forms of environmental deterioration (Mayer, 1967).

As a natural scientist it would not occur to me that in many cultures it is important to save face and prove virility by producing a child, as soon after marriage as possible. As a result of this, population planners must evidently aim at delaying the age of marriage or spreading out the production of children after the first. After it is brought to one's attention, it is easy to see that a tradition to produce many children would develop under conditions where one wants descendants and where few children survive to reach maturity (Sheps and Ridley, 1965).

In a Moslem country like Pakistan

where women will not be examined by a male physician, birth control by such measures as the intra-uterine device (IUD) is impracticable, and it is difficult to convey a monthly schedule of pill-taking to the poorly educated. However, just as the reproductive cycles of cattle can be synchronized by hormone treatments so that many cows can undergo simultaneous artificial insemination, so the menstrual cycles of populations of women can be synchronized. Then the instructions for contraception can take such a simple form as: "take a pill every night the moon shines." But in a country like Puerto Rico, the efforts of an aroused clergy to instill guilt feelings about the decision a woman must make each day can render the Pill ineffective. Here the IUD, which requires only one decision, is more practicable.

In any case, there is ample evidence that people the world over want fertility control. Voluntary sterilization is popular in India, Japan, and Latin America. In Japan and western European countries that have made legal abortion available upon request, the birth rates have fallen dramatically. With such recent techniques as the Pill and the IUD, and the impending availability of antimeiotic drugs which inhibit sperm production in the male, and anti-implantation drugs which can prevent pregnancy when taken as much as 3 days after exposure, practicable fertility control is at last available.

Prospects

I shall try to end on a note of optimism. Japan has shown that a determined people can in one generation bring the problem of excessive population growth under control. Japan's population is still growing slowly but they know how to stop it and, as the age distribution of her population adjusts to the new schedule of fertility, there may be no need for further deliberate actions.

Kingsley Davis (1967) has recently expressed skepticism about schemes for family planning on the grounds that they do not actually represent population policy but merely permit couples to determine their family size voluntarily. This is certainly true, but the evidence is overwhelming that a large percentage of the children born into the world today are unwanted. I think we must start by preventing these unwanted births and then take stock of what additional measures such as negative dependency allowances may be called for.

In the meantime, I consider it urgent that social and natural scientists get together and try to decide what an optimum size for the human population of the earth would be. If this can be achieved before some miscalculation, or noncalculation, sends the earth's environment into an irreversible decline, there is hope that the world can be saved. It is encouraging that the Soviet Union seems finally to have abandoned the dogma that overpopulation problems are by-products of capitalism and could not exist in a socialistic country (Cook, 1967). It is to be hoped that this will open the door to true international cooperation, possibly even within the United Nations, to try to solve man's most desperate problem.

REFERENCES

Belter, W.G., and D.W. Pearce. 1965. Radioactive waste management. *Reactor Tech.* (AEC), Jan. 1965, p. 203.

Berkner, L.V. 1966. Man versus technology. *Population Bull.*, **22:** 83–94.

Berkner, L.V., and L.C. Marshall. 1965. On the origin and rise of oxygen concentration in the earth's atmosphere. *J. Atmos. Sci.*, **22:** 225–261.

Bolin, B., and E. Eriksson. 1959. Changes in the carbon dioxide content of the atmosphere and sea due to fossil fuel consumption. In: *The Atmosphere and the Sea in Motion.* B. Bolin, Ed. The Rockefeller Institute Press, New York.

Borgstrom, G. 1965. *The Hungry Planet.* Macmillan, New York.

Clarke, F.W. 1920. The data of geochemistry. *U.S. Geol. Surv. Bull.*, 695.

Cook, R.C. 1967. Soviet population theory from Marx to Kosygin. *Population Bull.*, **23:** 85–115.

Davis, K. 1967. Population policy: will current programs succeed? *Science*, **158:** 730–739.

Deshler, W. 1967. *Sci. Am.*, **217** (4): 8.

Frank, W.J. 1964. Characteristics of nuclear explosives. In: *Engineering with Nuclear Explosives.* U.S. Atomic Energy Commission. TID-7695.

Graves, E. 1964. Nuclear excavation of a sea-level, Isthmian canal. In: *Engineering with Nuclear Explosives.* U.S. Atomic Energy Commission. TID-7695.

Linton, R.M. (Chairman). 1967. A strategy for a livable environment. Report of The Task Force on Environmental Health and Related Problems. U.S. Dept. H.E.W. Govt. Printing Office, Washington, D.C.

Lowdermilk, W.C. 1948. Conquest of the land through seven thousand years. USDA, Soil Cons. Serv. MP-32.

Mayer, J. 1967. Nutrition and civilization. *Trans. N.Y. Acad. Sci.*, **29:** 1014–1032.

Nye, P.H., and D.J. Greenland. 1960. The soil under shifting cultivation. Commonwealth Ag. Bur. Farnham Royal, Bucks, England. Tech. Comm. 51.

Parker, F.L. 1968. Wastes from fusion reactors. *Science*, **159:** 83.

Parker. F.L., and D.J. Rose. 1968. Wastes from fusion reactors. *Science*, **159:** 1376.

Parsons, J.J., and W.M. Denevan. 1967. Pre-Columbian ridged fields. *Sci. Am.*, **217**(1): 92–100.

Rubey, W.W. 1955. Development of the hydrosphere and atmosphere, with special reference to probable composition of the early atmosphere. Geol. Soc. Am., Spec. Paper 62, Crust of the earth, 631–650.

Sheps, M.C., and J.C. Ridley (Eds.), 1965. *Public Health and Population Change.* University of Pittsburgh Press.

Snow, J.A. 1967. Radioactive waste from reactors. *Scientist and Citizen*, **9:** 89–96.

Stead, F.W. 1964. Distribution in groundwater of radionuclides from underground nuclear explosions. In: *Engineering with Nuclear Explosives.* U.S. Atomic Energy Commission. TID-7695.

Tukey, J.W. (Chairman). 1965. Restoring the quality of our environment. Report of the President's Scientific Advisory Committee. Govt. Printing Office, Washington, D.C.

Urey, H.C. 1959. Primitive planetary atmospheres and the origin of life. In: *The Origin of Life on the Earth.* Macmillan, New York.

2.

Man Versus Nature

Peter Farb

Since boyhood, Peter Farb has been sensitive to and knowledgeable about the relationships among living things. He is a member of the Ecological Society, former secretary of the New York Entomological Society, and a fellow of the American Association for the Advancement of Science.

He has written widely on nature subjects. He is the author of many magazine articles and two very popular books: *Living Earth*, which concerns the ecology of the soil, and *Face of North America*. Peter Farb is also editor and coauthor of the North American Nature Series, which concern themselves with the ecology on the primary regions of the continent.

The despair felt by many people as they look upon a world drastically altered by the hand of man was rarely more deeply expressed than in the lines by Thomas Beddoes, an English poet of the last century:

Nature's polluted,
There's man in every secret corner of her
Doing damned wicked deeds. Thou art, old world
A hoary, atheistic, murdering star.

And truly, modern man, with his atomic gadgetry and penchant for uprooting the earth, does seem to have severed his intimate ties with the natural world. It is this unnatural break that makes many a modern man hanker for a primitive and simpler past, or for the remote civilizations that still persist in close harmony with nature. It is indeed true that primitive man more fully, although unknowingly, used ecological principles in his everyday life. He usually did not make the distinction modern man does—between himself and the surrounding natural world, as if all else in nature outside of man's body could be lumped together. Although the word "totem" comes from the language of the Ojibway Indians, who lived around the Great Lakes, the idea of totemism is found in primitive cultures around the world. Totemism generally means a sense of kinship between a group of people and a particular object or form of life, whether it be the parrot fish totem of an aboriginal clan in Australia or the tiger of a clan in India. Perhaps the idea of totemism persists unconsciously in this more complicated modern world, for its modern relic may be the symbols of today's nations, invariably an animal supposed to possess enviable attributes—the United States eagle, the Russian bear, the British lion.

Many supposedly primitive peoples do possess an impressive body of ecological insight. The South American Indians of the equatorial forest know their environment intimately—the places frequented by every kind of animal and the best ways to catch it, the names of the different trees and their attributes; the Masai of East Africa have been aware for centuries that mosquito bites cause malaria; the Eskimo long ago discovered that his sledge dogs were susceptible to the diseases of the wild arctic foxes. The Buddhist tradition in Asia protects animals because Buddha revered all forms of life and would not tolerate cruelty; as a result, the lands surrounding Buddhist monasteries in many cases have become virtual wildlife sanctuaries. An extreme of this tradition is Jainism: the Jain monk, bound to respect even the lives of insects, carries a broom to sweep them aside lest he tread on them accidently. That this kinship with the world of nature has persisted, despite the attempts of Western man to extirpate it, was demonstrated when attempts were made to control rats in warehouses in India; the natives placed in charge persisted in leaving water for the rats to drink. It is easy for Western men to ridicule these beliefs as sentimental or even foolish. Yet the great modern theologian and humanist, Dr. Albert Schweitzer, echoes this essential reverence for all living things: "The great fault of all ethics hitherto has been that they believed themselves to have to deal only with the relation of man to man."

Where then did man go astray in cleaving the ties of his heritage with the rest of nature? Although no one knows for sure, the process probably began in Asia Minor a scant 10,000 years ago—a mere 375 human generations—during the Neolithic, also called the New Stone Age because improved stone tools were then used by man. The preneolithic peoples, although few in number, interfered with their environment on a scale out of proportion to their scarcity. Some of their hunting methods were sheer waste, such as driving herds of animals over precipices as an easy way of killing them. Since his population was low and he was constantly on the move, however, man's influence on the environment was local and only temporary—giving the land time to recover until man came that way again. The onset of the Neolithic was probably not a sudden economic revolution; rather agriculture and domestication of animals must have emerged gradually out of the Mesolithic. But these tendencies gathered momentum in the Neolithic; it was then that man clearly emerged as the only animal that set out to subdue his environment instead of adapting to it.

Neolithic man interfered with nature by growing crops and turning animals to his use—thus inevitably destroying forests, causing soil to wash away, polluting rivers with sediment, accelerating the natural and gradual processes of erosion and plant succession. To prepare the fields and harvest the crops, to care for the animals, to process the foodstuffs and animal byproducts, required a communal life of closely packed villages rather than small nomadic tribes. As the intricacy of the village system grew, it was soon found that one village might have a good supply of stone for axes, another of clay for pottery, a third of plant fibers for baskets. These products were distributed by trade, which necessitated making roads. It was only a logical extension of these tendencies that has resulted in today's complex of metropolises, the destruction of primeval forests, our stress-causing social patterns—and a modern view of nature as something to be subjugated, not as an abundance with which to live in harmony.

The neolithic way of life persists today, most conspicuously in the headwaters of the impenetrable Amazonian forests, in the interior of New Guinea, among the Berber tribesmen of North Africa. Somewhat less conspicuously, the Neolithic still exists even in the so-called developed countries, where farm lands cluster around a village. Man still cuts forests haphazardly and plows deep into natural grasslands, and in many even rather advanced places his faith in wonder cures and religious healing differs little from a primitive faith in the shaman who used herbs and magic to cure human diseases. "Neolithic culture is much more than a subject of inquiry by prehistorians," states the anthropologist Carlton S. Coon. "Moving out of it may be the world's most difficult problem."

Ever since the Neolithic, man has had the power—and the responsibility—of a destiny ascribed to him in *Genesis*: to "have dominion over all the earth." But during his dominion he has broken nearly every ecological principle of energy flow, isolation, community interaction and population control—and so far he seems to have gotten away with it. He has misused land, forests, water, sea fisheries; he has spread diseases and alien forms of life, at the same time extirpating native forms; in recent centuries he has been reproducing like the lemming and introducing too many of his own kind into ecosystems that cannot hold them. The fact that man's present state in the world is adjudged perilous can be seen by the hair-raising titles of some books published in the last few decades—written not by cranks but by scientists and humanists deeply concerned about the human conditions—*Road to Survival, The Rape of the Earth, Our Plundered Planet, The Geography of Hunger, The Limits of the Earth, The Prevalence of People*, and many others.

In fact, by the 20th century, man had finally conquered the biosphere and colonized the earth. His domination is due to his brain, and appropriately it has been suggested by one scientist that the new state of the biosphere be referred to as the noösphere—derived from the Greek word for "mind."

Modern man must soon make a decision. Either he will abide by technology and live in a man-centered planet in which he becomes increasingly aloof from several billion years of biological experiences—or he will work in harmony with the principles of ecology and use for himself the same criteria that apply to plants and other animals. The choice is not a simple one. The engineer assures us that he can control nature, the ecologist states that he can live in harmony with it—and both views can be sustained by argument. But the viewpoint of this book has been that man belongs *to* nature and cannot long remain separate from various biological laws without an eventual day of reckoning—that he must develop what the American conservationist Aldo Leopold called an "ecological conscience." Following the ecological viewpoint, let us see some of the primary principles that man is challenging, and their possible consequences.

In the first place, man has markedly altered the physical environment itself—as do all living things. But whereas the changes in environment promoted by other living things encourage renewal and turnover, man has in many cases disrupted the opportunities for rebirth. While man remained simply a food gatherer, he caused little modification of the environment, and the turnover was rapid. At this stage, he might be considered as a commensal with his environment as a whole. The Indians of the Great Plains of North America, for example, fitted in quite well with the grassland community. They were high-level consumers of the bison and did not much upset the intricate relationship between producing grasses and the primary consumers, the bison, or the wolves that preyed upon them. However, as soon as man reached the level of domestication and agriculture, he began to change the environment physically. He struggled against the regrowth of natural vegetation and thus prevented the successional stages of vegetation to the climax. In planting crops and breeding animals, he made certain species dependent upon him for their survival—wheat cannot now procreate by itself, and a cow abandoned on the plains would in a short time fall prey to coyotes. Man has converted the cow from a wild animal, well able to defend itself and its young, into a walking milk factory.

As man progressed to the next level, that of industry, he went even further—he created a whole new ecosystem and substituted it for the natural one. A dam thrown across a river alters the entire natural drainage pattern and thereby upsets the water tables of the area. The toxic fumes from smelters have utterly destroyed all life for miles around them, creating such man-made deserts as those once caused in southeastern Tennessee by copper mining and at Sudbury, Ontario, by nickel fumes. In the latest state—urbanization—man has completely replaced all of the natural elements in the environment with artificial ones. Instead of soil, natural water systems, plant and animal communities, there now exist concrete surfaces, sewage pipes, and parks composed largely of alien plants and animals. In urban centers man has virtually suppressed all life that ever existed there and has substituted a community of little variety that repeats itself in cities around the world. Usually the only animals that live with man in this urban environment are his domesticated dogs, cats, goldfish, turtles, canaries, budgies and a few other pets. The birds found in the North American urban environment are either aliens—street pigeons, starlings, house sparrows—or the few natives that have been able to adapt to urban living in parks of foreign vegatation. In the cities there also live man's unwelcome congeners: rats, houseflies, cockroaches, lice and numerous microorganisms. During the various steps up to urbanization, man has so radically altered the face of the earth—the very physical environment to which life had gradually become adapted—that any repopulation by the original plants and animals is remote indeed.

Hand in hand with the alteration of the physical environment went the disruption of the original community. A natural community in an area that has not seen sudden geologic change is a tight fabric of food webs which are connected with the environment. Man has simplified and rearranged the energy relationship that once prevailed in these communities; he has eliminated complicating threads from the fabric of life to such an extent that in many places it is threadbare indeed. Primitive man was a second- or third-order consumer—he lived off other animals that had first consumed the plant producer, or off the predatory animals that consumed the herbivores. When he gathered fruits and nuts he was a first-order consumer, but with the advent of agriculture he broadened his food sources, substituting edible plants for inedible pine needles and oak leaves. He simplified natural grasslands by growing wheat, corn, potatoes, cassava and other crops in a single-crop system. To meet his demands for wood, he replaced the luxuriant deciduous forests with quick-growing conifers. He satisfied his taste for meat by growing food for his domesticated herbivores, such as sheep and cattle. To make himself secure as a first-order consumer, he not only had to shorten the natural food chains but also eliminate whatever predators he believed to be in conflict with his interests.

Because insects consume his food crops and his ornamental flowers as well, man has broadcast upon the landscape insecticides that eliminate not only his direct competitors, but many beneficial and abounding noncompetitive forms of life as well. Without quite realizing it, he has thus left himself vulnerable. For this kind of simplified biological community—such as exists naturally on certain remote islands or in the tundra—can be upset much more easily when one of its parts gets out of kilter than can a diversified community. Wheat stretching as far as the eye can see undoubtedly can be harvested more efficiently by machines; but a single crop of wheat is also susceptible to sudden onslaughts of insects or fungi which can build up their populations catastrophically in a short time. Man is constantly on guard against creatures he thought he had subdued, for uneasy lies the crown of husbandry where even a small population of potential usurpers still exists. The only way man has been able to produce successful wheat crops is by always staying one small step ahead of the blight of rust—by turning out new varieties of wheat faster than the fungus can adapt to them. And he has found that one peculiar fact about the use of insecticides is that they actually encourage other pests. In recent years DDT has been applied widely in orchards around the world for the control of certain pests, among them red mites that attack fruit. But instead of killing red mites, DDT kills its enemies without much harming the mite itself. As a result, there has arisen the new problem of a worldwide abundance of red mites in orchards.

There is no denying that in general the man-made community works efficiently—though it may continue to do so for only a limited time. Nevertheless, this efficiency has allowed man to circumvent another set of ecological principles, those governing populations.

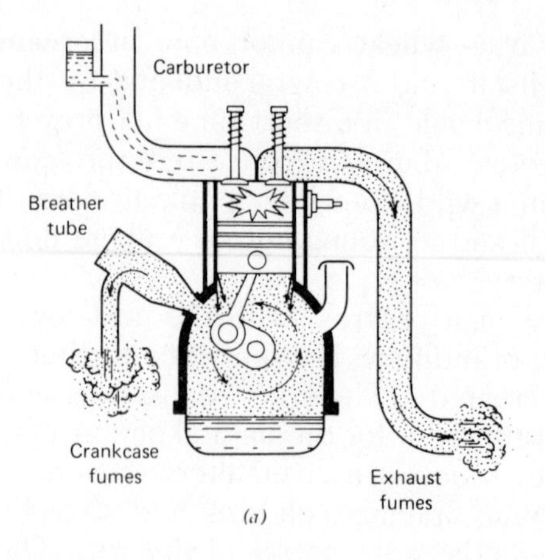

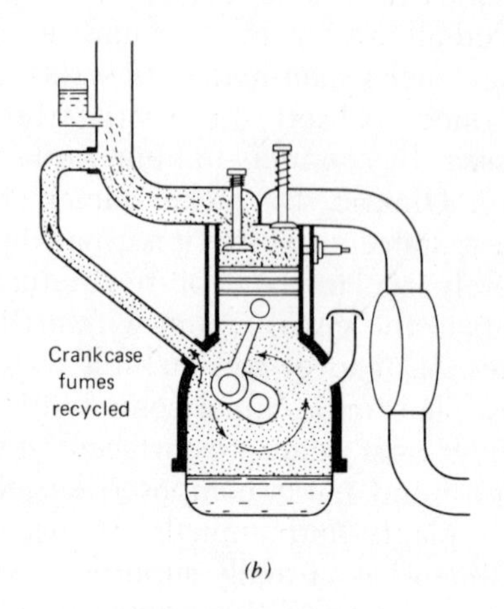

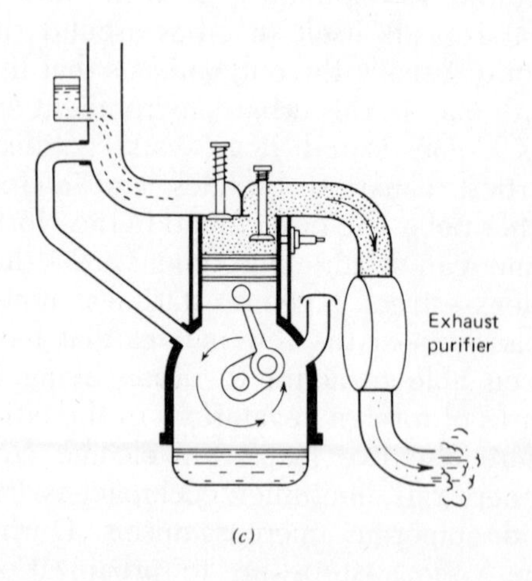

The Deadly Auto Engine. Of all the various agents of air pollution, the gasoline engines powering the world's millions of automobiles are among the most difficult to control. The toxic, unburned hydrocarbons in crankcase fumes (*a*) could be fairly easily (and cheaply: $5 to $10) eliminated by a "blowby" recycler, which feeds them back from the breather tube into the caburetor (*b*) so that they are burned in the engine cylinders. More complicated are systems to detoxify the exhaust fumes themselves (*c*): these may employ either a catalyst for low-temperature (200° to 1,500°F.) combustion, or an "after-burner" which, like a blowtorch, burns the hydrocarbons at temperatures around 2,800°F. The real problem: to win approval of such devices by car owners.

Today there are some three billion human beings on the planet. About 270,000 infants are born every day, and about 142,000 people die every day—making a surplus of some 128,000 daily, or a population increase each month equivalent to that of Chicago. When Christ was born there were probably between 250 and 350 million people on earth. It was not until 1650 that this number doubled; in only 200 years after that it doubled again; and in the mere hundred years between 1850 and 1950 it doubled once more. Unless one believes that man has utterly obliterated the biosphere and substituted for it a successful noösphere, then this sort of population increase cannot continue much longer. If the annual rate of human growth were to continue at its present 1.8 per cent, by 1980 there would be close to another billion people—and not long after the year 2000 the world population would double today's.

Only during the relatively last few seconds of man's short tenure on earth has the human population increase become a problem. The maximum worldwide human population during paleolithic times probably did not exceed 10 million; this number started to rise with the coming of the neolithic revolution when agriculture made it possible for the land to support a higher population. One of the first people to become concerned was Thomas Malthus, who published his *Essay on Population* in 1798; at that time the human population was less than a third of what it is today. Malthus' approach was a sound ecological one. He calculated that human populations possess the biotic potential to double every 25 years but that food resources do not multiply so rapidly. Therefore, human population, unless held in check, will increase up to the limit of its food supply. The only logical answer to what he called this "dismal thorem" was human misery. It was fashionable to deride Malthus during the optimism that swept the 19th Century, and there are still optimists today who state that Malthus was wrong. The fact is that Malthus has been proven correct more than once: in Ireland in the early 19th century, the introduction of the potato, a cheap and easily raised staple food, brought on a population explosion that was finally halted by the potato blight and resulting famine in 1845. In India today, constant famine graphically demonstrates the inability of the environment to keep pace with human growth. Some 10,000 people around the world die every day of starvation or malnutrition. Despite agricultural advances, irrigation and reclamation of deserts, more than half of the world's population still lives in perpetual hunger. Only the Western civilizations have seemingly escaped the workings of the dismal theorem by their present agricultural abundance. However, Malthus has not been disproved by Western technology; there merely has not been time as yet for the truth of his dismal theorem to be tested.

The entire question of the future size of human populations is enveloped in unknowns. No one is certain how far future technology can increase the carrying capacity of the planet for man, nor can one be sure that future population growth will not exceed even the advances of technology. In populations of other animals, there are controls built into the ecosystem in the form of competition, starvation, predation and disease. But man rightly does everything in his power to prevent these ecological controls from operating upon the human population and causing human suffering. He tries to keep diseases from creating violent population fluctuations such as occurred during the Middle Ages when about a fourth of Europe's population died from the Black Death. He has long ago eliminated predation by other animals as a factor in human population control. He has made almost

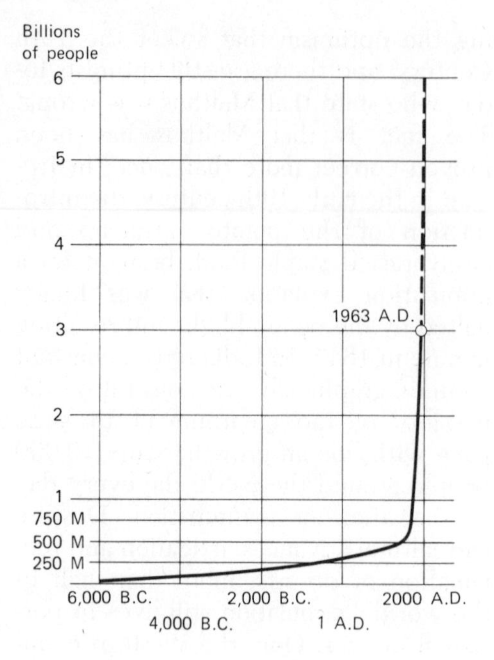

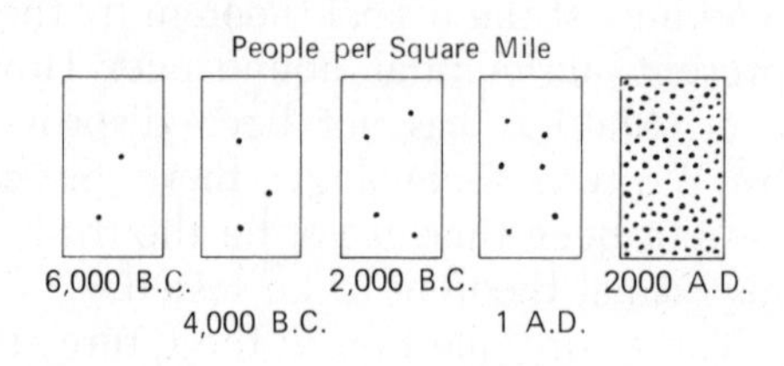

The Population Explosion. The graph shown here traces the gradual increase in human numbers over thousands of years, leading up to the vast explosion which is currently under way. In the last century the population has doubled and in another half century it may double again (dotted line), conjuring up the frightening image of 125 individuals per square mile of the earth, compared to about six per square mile in 1 A.D. Even now the overload is worse than these averages indicate. Only one tenth of the earth is arable, and most of the croplands lie outside of Asia —yet half of the world's people are concentrated there.

continuous attempts to eliminate competition within his own species by trying to outlaw war. With the checks upon human populations largely removed, man has increasingly relied upon new technology to provide for his constantly growing numbers. But the technological opportunities—controlling the climate, growing algae for food in artificial ponds, desalting the seas—all have their limits and at best they can only buy time until man works out some way of controlling his own numbers by himself.

One of the startling revelations of the last United States census was that Vermont had not increased its population, a fact that led an eminent Vermonter, the poet Robert Frost, to say that he was glad it had not. "We want to grow right," he stated. Growing right means that man should possess the amenities of life in the form of open space, a diet of meat rather than algae, of water for recreation rather than solely for efficient fish farming. Among these amenities is the one of living on a planet sufficiently uncrowded for other forms of life to exist. The grizzly bear has been hounded nearly to extinction and a mere 1,000 survive in the United States proper, plus 5,000 more in remote areas of Alaska. There is something that strikes at the ecological conscience here; it is not sufficient that the grizzly should survive only as a relict species in the distant north. "Relegating grizzlies to Alaska is about like relegating happiness to heaven," wrote Aldo Leopold, "one may never get there."

At this time man simply does not have the necessary information to make sound judgments about his future; ironically, he knows less about the laws governing the rises and falls of his own populations than he does about those of many other kinds of animals. But fever, famine and war are still the three basic controls upon human numbers. Despite man's best efforts, war looms as an everyday possibility, and hunger is a reality to half of the world. Seemingly, the control of infectious diseases marks man's greatest success in eliminating the checks upon his numbers. But man's discovery of the microbial world and his attempts to control it have brought with it ecological disruptions. Although microbes have not yet been used as weapons of war, they have been used in what is known as the "biological control" of species that man regards as pests. In Australia in 1950 and thereafter in western Europe, the virus of myxomatosis from Brazil decreased the rabbit hordes without harm to other species. The unleashing of insect pathogens against insect pests has been gaining momentum in recent years and, if used wisely, can be an ecological alternative to the widespread spraying of insecticides. But there is also danger in spreading these pathogens on the landscape, for this exposes whole populations of animals to the impact of strains of pathogens that did not evolve and reach their present density by natural processes.

Until the 1945 atomic blast at Alamogordo, no animal in a natural environment had been exposed to the effects of radiation beyond the low background level that always has been part of the terrestrial surroundings. In recent years man has released radiation into the environment in tremendous amounts, with effects upon individuals, populations, communities and ecosystems that cannot yet be assessed. Radioactive substances raining down upon the landscape after an atomic test penetrate the soils and water; many organisms absorb them and concentrate them in their tissues. So far, the dilute amounts found in the environment are probably of no serious consequence, but there are grounds for deep concern. Very little is known about the exact effects of low-level, long-term exposure to radiation upon ecosystems, but facts are available about the effect of radioactivity due to X-rays. One fact stands out clearly: there are substantial differences among organisms in their ability to tolerate massive doses. Ten thousand rads—units of absorbed dose—may change the rate at which bacteria in a culture divide, but it might take more than a million

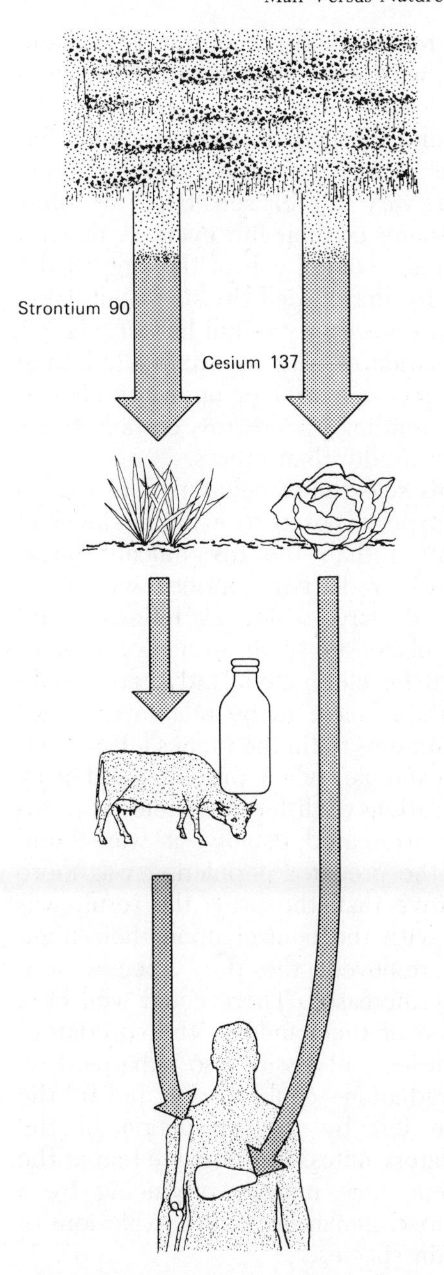

The Paths of Fallout. The core of the fallout problem lies not so much in the quantity of fallout but in the way certain long-lived radioactive particles become concentrated in the course of a food chain. Thus the radioisotope strontium 90 is absorbed by cattle as they eat tainted grass; it is passed on to man in milk and cheese, and, like calcium, finally concentrates in the bone marrow, where it may lead to leukemia or bone cancer. Cesium 137, less soluble, is picked up directly from vegetables and concentrates in soft tissues such as the liver or gonads, where it may be a menace to the genes, bearers of heredity.

rads to kill all the bacteria. The mammals, which are generally more susceptible, can be affected by doses as low as 100 and killed by 1,000 rads. But even these figures are misleading, for living things vary in their sensitivity at different stages of their life cycle. A dose of 163 rads will kill half of the eggs of the fruit fly during cell division—but 100,000 are necessary to kill half of a larger population of fruit fly adults. Radiation thus acts as a strainer upon the ecosystem, tending to destroy certain types more rapidly than others.

This selectivity means that if a community is exposed to a higher level of radiation than the insignificant background radiation under which it evolved, large-scale disruptions could take place. Sensitive strains or species would be eliminated, and there would inevitably have to be adaptations and adjustments to fill the niches left vacant. For example, when predator and prey populations of different species of mites were irradiated equally, it was found that the predator population was more sensitive than the prey; the result was that with the control upon their numbers removed, the prey species suddenly increased. There could well be a plague of them unless other predatory species—of necessity also unharmed by the radiation—could step in and fill the niche left by the irradiation of the predatory mites. It is possible that at the present time man is producing by a radiation similar ecological explosions of certain species.

Even if small doses of radioisotopes are introduced into the air or water, they may undergo unexpected concentrations, with far-reaching effects on food chains—an individual high in the food chain may then represent a lethal package to the final consumer. This was demonstrated at the Hanford atomic power plant on the Columbia River in Washington, where minute amounts of various isotopes were released into the water at a constant rate, making it possible to measure the way in which radioactivity might become concentrated in food chains under natural conditions. One of the measurements made by the Hanford scientists revealed that the eggs of ducks and geese contained radioactive phosphorus in concentrations averaging 200,000 times higher than the solution released into the water—and occasionally the concentration soared as high as 1,500,000 times. In another case, of air contamination, iodine became concentrated in the thyroids of jack rabbits at an average level 500 times as great as released. These levels may not yet be critical for the organisms concerned; despite the high concentrations of phosphorus in the waterfowl eggs, the eggs still hatched. There is, however, little knowledge of the genetic damage caused to the species, although it certainly exists. If present pollution of the environment by radioactivity persists, a threshold will soon be reached beyond which serious damage will be done to the ecosystem by the elimination of sensitive organisms. Even barring nuclear holocaust and further testing of atomic bombs, man's peaceful uses of atomic energy alone will create a garbage disposal problem of the nuclear wastes into the environment.

As man looks about him at the noösphere he has created, he is tempted to call a moratorium on technology and give himself the opportunity for needed soul searching, hopefully to come up with the answers to problems that threatened his very survival. But stopping the clock does not solve ecological problems any more than closing the banks solves economic ones. In ecology there are no Gardens of Eden, no Utopias, no ways of turning the clock back to a simpler existence. It is, in fact, probably of the essence of the human animal that he consciously pursues change. This has set him apart from the rest of the natural world by leading him to abandon the tropical environment to which he was biologically adapted, and has brought him ultimately to the changes he has wrought during his bid for control of nature. Charles Darwin in *The Descent of Man* recognized the human need for change and its biological implications: "If all our women were to become as beautiful as the Venus de' Medici, we should for a time be charmed; but we should soon wish for variety; and as soon as we had obtained variety, we should wish to see certain characters a little exaggerated." Man's estate seemingly is to live with unsatisfied aspirations; it is what has led to his present predicament of being a ruler over the earth without knowing the rules.

However, these rules do exist in an ecological view of the world and of man's place in it. Man may be able to harness the power of the atom and make a loud noise and to send space vehicles whirling around the globe. But he has not revoked any of the physical laws which govern the universe—he must still use an old-fashioned parachute to counteract the law of gravity as his space capsule lands. Similarly, although man can shave a forest and dam a stream, he has not repealed the laws that have governed the procession of life upon the planet. He cannot for long toy with the rules that govern ecosystems and communities, that provide built-in balances in the diversity of life around him. Of all the principles of ecology, the primary one was stated in the 17th Century by the philosopher Francis Bacon: "We cannot command nature except by obeying her."

3.

The Politics of Ecology: The Question of Survival

Aldous Huxley

Aldous Huxley is the author of some of the most challenging and beautifully written novels in the English language. *Point Counter Point, Brave New World,* and *Ape and Essence* are the most famous. His essays have been printed in most languages and encompass many areas of human thought. This eloquent essay faces the sometimes jarring realities of overpopulation and considers some of the possible solutions for this problem.

In politics, the central and fundamental problem is the problem of power. Who is to exercise power? And by what means, by what authority, with what purpose in view, and under what controls? Yes, under what controls? For, as history has made it abundantly clear, to possess power is *ipso facto* to be tempted to abuse it. In mere self-preservation we must create and maintain institutions that make it difficult for the powerful to be led into those temptations which, succumbed to, transform them into tyrants at home and imperialists abroad.

For this purpose what kind of institutions are effective? And, having created them, how can we guarantee them against obsolescence? Circumstances change, and, as they change, the old, the once so admirably effective devices for controlling power cease to be adequate. What then? Specifically, when advancing science and acceleratingly progressive technology alter man's long-established relationship with the planet on which he lives, revolutionize his societies, and at the same time equip his rulers with new and immensely more powerful instruments of domination, what ought we to do? What *can* we do?

Very briefly let us review the situation in which we now find ourselves and, in the light of present facts, hazard a few guesses about the future.

On the biological level, advancing science and technology have set going a revolutionary process that seems to be destined for the next century at least, perhaps for much longer, to exercise a decisive influence upon the destinies of all human societies and their individual members. In the course of the last fifty years extremely effective methods for lowering the prevailing rates of infant and adult mortality were developed by Western scientists. These methods were very simple and could be applied with the expenditure of very little money by very small numbers of not very highly trained technicians. For these reasons, and because everyone regards life as intrinsically good and death as intrinsically bad, they were in fact applied on a world-wide scale. The results were spectacular. In the past, high birth rates were balanced by high death rates. Thanks to science, death rates have been halved but, except in the most highly industrialized, contraceptive-using countries, birth rates remain as high as ever. An enormous and accelerating increase in human numbers has been the inevitable consequence.

At the beginning of the Christian era,

Reprinted by permission of *The Center Magazine*, a publication of the Center for the Study of Democratic Institutions, and The Fund for the Republic. From *The Politics of Ecology: The Question of Survival*, by Aldous Huxley, copyright 1963.

so demographers assure us, our planet supported a human population of about two hundred and fifty millions. When the Pilgrim Fathers stepped ashore, the figure had risen to about five hundred millions. We see, then, that in the relatively recent past it took sixteen hundred years for the human species to double its numbers. Today world population stands at three thousand millions. By the year 2000, unless something appallingly bad or miraculously good should happen in the interval, six thousand millions of us will be sitting down to breakfast every morning. In a word, twelve times as many people are destined to double their numbers in one-fortieth of the time.

This is not the whole story. In many areas of the world human numbers are increasing at a rate much higher than the average for the whole species. In India, for example, the rate of increase is now 2.3 per cent per annum. By 1990 its four hundred and fifty million inhabitants will have become nine hundred million inhabitants. A comparable rate of increase will raise the population of China to the billion mark by 1980. In Ceylon, in Egypt, in many of the countries of South and Central America, human numbers are increasing at an annual rate of 3 per cent. The result will be a doubling of their present populations in approximately twenty-three years.

On the social, political, and economic levels, what is likely to happen is an underdeveloped country whose people double themselves in a single generation, or even less? An underdeveloped society is a society without adequate capital resources (for capital is what is left over after primary needs have been satisfied, and in underdeveloped countries most people never satisfy their primary needs); a society without a sufficient force of trained teachers, administrators, and technicians; a society with few or no industries and few or no developed sources of industrial power; a society, finally, with enormous arrears to be made good in food production, education, road building, housing, and sanitation. A quarter of a century from now, when there will be twice as many of them as there are today, what is the likelihood that the members of such a society will be better fed, housed, clothed, and schooled than at present? And what are the chances in such a society for the maintenance, if they already exist, or the creation, if they do not exist, of democratic institutions?

Not long ago Mr. Eugene Black, the former president of the World Bank, expressed the opinion that it would be extremely difficult, perhaps even impossible, for an underdeveloped country with a very rapid rate of population increase to achieve full industrialization. All its resources, he pointed out, would be absorbed year by year in the task of supplying, or not quite supplying, the primary needs of its new members. Merely to stand still, to maintain its current subhumanly inadequate standard of living, will require hard work and the expenditure of all the nation's available capital. Available capital may be increased by loans and gifts from abroad; but in a world where the industrialized nations are involved in power politics and an increasingly expensive armament race, there will never be enough foreign aid to make much difference. And even if the loans and gifts to underdeveloped countries were to be substantially increased, any resulting gains would be largely nullified by the uncontrolled population explosion.

The situation of these nations with such rapidly increasing populations reminds one of Lewis Carroll's parable in *Through the Looking Glass*, where Alice and the Red Queen start running at full speed and run for a long time until Alice is completely out of breath. When they stop, Alice is amazed to see that they are still at their starting point. In the looking glass world, if you wish to retain your present position, you must run as fast as you can. If you wish to get ahead, you must run at least twice as fast as you can.

If Mr. Black is correct (and there are plenty of economists and demographers who share his opinion), the outlook for most of the world's newly independent and economically non-viable nations is gloomy indeed. To those that have shall be given. Within the next ten or twenty years, if war can be avoided, poverty will almost have disappeared from the highly industrialized and contraceptive-using societies of the West. Meanwhile, in the underdeveloped and uncontrolledly breeding societies of Asia, Africa, and Latin America the condition of the masses (twice as numerous, a generation from now, as they are today) will have become no better and may even be decidedly worse than it is at present. Such a decline is foreshadowed by current statistics of the Food and Agriculture Organization of the United Nations. In some underdeveloped regions of the world, we are told, people are somewhat less adequately fed, clothed, and housed than were their parents and grandparents thirty and forty years ago. And what of elementary education? UNESCO recently provided an answer. Since the end of World War II heroic efforts have been made to teach the whole world how to read. The population explosion has largely stultified these efforts. The absolute number of illiterates is greater now than at any time.

The contraceptive revolution which, thanks to advancing science and technology, has made it possible for the highly developed societies of the West to offset the consequences of death control by a planned control of births, has had as yet no effect upon the family life of people in underdeveloped countries. This is not surprising. Death control, as I have already remarked, is easy, cheap, and can be carried out by a small force of technicians. Birth control, on the other hand, is rather expensive, involves the whole adult population, and demands of those who practice it a good deal of forethought and directed will-power. To persuade hundreds of millions of men and women to abandon their tradition-hallowed views of sexual morality, then to distribute and teach them to make use of contraceptive devices or fertility-controlling drugs—this is a huge and difficult task, so huge and so difficult that it seems very unlikely that it can be successfully carried out, within a sufficiently short space of time, in any of the countries where control of the birth rate is most urgently needed.

Extreme poverty, when combined with ignorance, breeds that lack of desire for better things which has been called "wantlessness"—the resigned acceptance of a subhuman lot. But extreme poverty, when it is combined with the knowledge that some societies are affluent, breeds envious desires and the expectation that these desires must of necessity, and very soon, be satisfied. By means of the mass media (those easily exportable products of advancing science and technology) some knowledge of what life is like in affluent societies has been widely disseminated throughout the world's underdeveloped regions. But, alas, the science and technology which have given the industrial West its cars, refrigerators, and contraceptives have given the people of Asia, Africa, and Latin America only movies and radio broadcasts, which they are too simple-minded to be able to criticize, together with a population explosion, which they are still too poor and too tradition-bound to be able to control by deliberate family planning.

In the context of a 3, or even of a mere 2 per cent annual increase in numbers, high expectations are foredoomed to disappointment. From disappointment, through resentful frustration, to widespread social unrest the road is short. Shorter still is the road from social unrest, through chaos, to dictatorship, possibly of the Communist party, more probably of generals and colonels. It would seem, then, that for two-thirds of the human race now suffering from the consequences of uncontrolled breeding in a context of industrial backwardness, poverty, and illiteracy, the prospects for democracy, during the next ten or twenty years, are very poor.

From underdeveloped societies and the probable political consequences of their explosive increase in numbers we now pass to the prospect for democracy in the fully industrialized, contraceptive-using societies of Europe and North America.

It used to be assumed that political freedom was a necessary pre-condition of scientific research. Ideological dogmatism and dictatorial institutions were supposed to be incompatible with the open-mindedness and the freedom of experimental action, in the absence of which discovery and invention are impossible. Recent history has proved these comforting assumptions to be completely unfounded. It was under Stalin that Russian scientists developed the A-bomb and, a few years later, the H-bomb. And it is under a more-than-Stalinist dictatorship that Chinese scientists are now in process of performing the same feat.

Another disquieting lesson of recent history is that, in a developing society, science and technology can be used exclusively for the enhancement of military power, not at all for the benefit of the masses. Russia has demonstrated, and China is now doing its best to demonstrate, that poverty and primitive conditions of life for the overwhelmingly majority of the population are perfectly compatible with the wholesale production of the most advanced and sophisticated military hardware. Indeed, it is by deliberately imposing poverty on the masses that the rulers of developing industrial nations are able to create the capital necessary for building an armament industry and maintaining a well equipped army, with which to play their parts in the suicidal game of international power politics.

We see, then, that democratic institutions and libertarian traditions are not at all necessary to the progress of science and technology, and that such progress does not of itself make for human betterment at home and peace abroad. Only where democratic institutions already exist, only where the masses can vote their rulers out of office and so compel them to pay attention to the popular will, are science and technology used for the benefit of the majority as well as for increasing the power of the State. Most human beings prefer peace to war, and practically all of them would rather be alive than dead. But in every part of the world men and women have been brought up to regard nationalism as axiomatic and war between nations as something cosmically ordained by the Nature of Things. Prisoners of their culture, the masses, even when they are free to vote, are inhibited by the fundamental postulates of the frame of reference within which they do their thinking and their feeling from decreeing an end to the collective paranoia that governs international relations. As for the world's ruling minorities, by the very fact of their power they are chained even more closely to the current system of ideas and the prevailing political customs; for this reason they are even less capable than their subjects of expressing the simple human preference for life and peace.

Some day, let us hope, rulers and ruled will break out of the cultural prison in which they are now confined. Some day . . . And may that day come soon! For, thanks to our rapidly advancing science and technology, we have very little time at our disposal. The river of change flows ever faster, and somewhere downstream, perhaps only a few years ahead, we shall come to the rapids, shall hear, louder and ever louder, the roaring of a cataract.

Modern war is a product of advancing science and technology. Conversely, advancing science and technology are products of modern war. It was in order to wage war more effectively that first the United States, then Britain and the USSR, financed the crash programs that resulted so quickly in the harnessing of atomic forces. Again, it was primarily for military purposes that the techniques of automation, which are now in process of revolutionizing industrial production and the whole system of administrative and bureaucratic control, were first developed. "During World War II," writes Mr. John Diebold, "the theory and use of feedback was studied in great detail by a number of scientists both in this country and in Britain. The introduction of rapidly moving aircraft very quickly made traditional gun-laying techniques of anti-aircraft warfare obsolete. As a result, a large part of scientific manpower in this country was directed towards the development of self-regulating devices and systems to control our military equipment. It is out of this work that the technology of automation as we understand it today has developed."

The headlong rapidity with which scientific and technological changes, with all their disturbing consequences in the fields of politics and social relations, are taking place is due in large measure to the fact that, both in the USA and the USSR, research in pure and applied science is lavishly financed by military planners whose first concern is in the development of bigger and better weapons in the shortest possible time. In the frantic effort, on one side of the Iron Curtain, to keep up with the Joneses—on the other, to keep up with the Ivanovs—these military planners spend gigantic sums on research and development. The military revolution advances under forced draft, and as it goes forward it initiates an uninterrupted succession of industrial, social, and political revolutions. It is against this background of chronic upheaval that the members of a species, biologically and historically adapted to a slowly changing environment, must now live out their bewildered lives.

Old-fashioned war was incompatible, while it was being waged, with democracy. Nuclear war, if it is ever waged, will prove in all likelihood to be incompatible with civilization, perhaps with human survival. Meanwhile, what of the preparations for nuclear war? If certain physicists and military planners had their way, democracy, where it exists, would be replaced by a system of regimentation centered upon the bomb shelter. The entire population would have to be systematically drilled in the ticklish operation of going underground at a moment's notice, systematically exercised in the art of living troglodytically under conditions resembling those in the hold of an eighteenth-century slave ship. The notion fills most of us with horror. But if we fail to break out of the ideological prison of our nationalistic and militaristic culture, we may find ourselves compelled by the military consequences of our science and technology to descend into the steel and concrete dungeons of total and totalitarian civil defense.

In the past, one of the most effective guarantees of liberty was governmental inefficiency. The spirit of tyranny was always willing; but its technical and organizational flesh was weak. Today the flesh is as strong as the spirit. Governmental organization is a fine art, based upon scientific principles and disposing of marvelously efficient equipment. Fifty years ago armed revolution still had some chance of success. In the context of modern weaponry a popular uprising is foredoomed. Crowds armed with rifles and home-made grenades are no match for tanks. And it is not only to its armament that a modern government owes its overwhelming power. It also possesses the strength of superior knowledge derived from its communication systems, its stores of accumulated data, its batteries of computers, its network of inspection and administration.

Where democratic institutions exist and the masses can vote their rulers out of office, the enormous powers with which science, technology, and the arts of organization have endowed the ruling minority are used with discretion and a decent regard for civil and political liberty. Where the masses can exercise no control over their rulers, these powers are used without compunction to enforce ideological orthodoxy and to strengthen the dictatorial state. The nature of science and technology is such that it is peculiarly easy for a dictatorial government to use them for its own anti-democratic purposes. Well financed, equipped and organized, an astonishingly small number of scientists and technologists can achieve prodigious results. The crash program that produced the A-bomb and ushered in a new historical era was planned and directed by some four thousand theoreticians, experimenters, and engineers. To parody the word of Winston Churchill, never have so many been so completely at the mercy of so few.

Throughout the nineteenth century the State was relatively feeble, and its interest in, and influence upon, scientific research were neglible. In our day the State is everywhere exceedingly powerful and a lavish patron of basic and *ad hoc* research. In Western Europe and North America the relations between the State and its scientists on the one hand and individual citizens, professional organizations, and industrial, commercial, and educational institutions on the other are fairly satisfactory. Advancing science, the population explosion, the armament race, and the steady increase and centralization of political and economic power are still compatible, in countries that have a libertarian tradition, with democratic forms of government. To maintain this compatibility in a rapidly changing world, bearing less and less resemblance to the world in which these democratic institutions were developed—this, quite obviously, is going to be increasingly difficult.

A rapid and accelerating population increase that will nullify the best efforts of underdeveloped societies to better their lot and will keep two-thirds of the human race in a condition of misery in anarchy or of misery under dictatorship, and the intensive preparations for a new kind of war that, if it breaks out, may bring irretrievable ruin to the one-third of the human race now living prosperously in highly industrialized societies—these are the two main threats to democracy now confronting us. Can these threats be eliminated? Or, if not eliminated, at least reduced?

My own view is that only by shifting our collective attention from the merely political to the basic biological aspects of the human situation can we hope to mitigate and shorten the time of troubles into which, it would seem, we are now moving. We cannot do without politics; but we can no longer afford to indulge in bad, unrealistic politics. To work for the survival of the species as a whole and for the actualization in the greatest possible number of individual men and women of their potentialities for good will, intelligence, and creativity—this, in the world of today, is good, realistic politics. To cultivate the religion of idolatrous nationalism, to subordinate the interests of the species and its individual members to the interests of a single national state and its ruling minority—in the context of the population explosion, missiles, and atomic warheads, this is bad and thoroughly unrealistic politics. Unfortunately, it is to bad and unrealistic politics that our rulers are now committed.

Ecology is the science of mutual relations of organisms with their environment and with one another. Only when we get it into our collective head that the basic problem confronting twentieth-century man is an ecological problem will our politics improve and become realistic. How does the human race propose to survive and, if possible, improve the lot and the intrinsic quality of its individual members? Do we propose to live on this planet in symbiotic harmony with our environment? Or, preferring to be wantonly stupid, shall we choose to live like murderous and suicidal parasites that kill their host and so destroy themselves?

Committing that sin of overweening bumptiousness, which the Greeks called *hubris*, we behave as though we were not members of earth's ecological community, as though we were privileged and, in some sort, supernatural beings and could throw our weight around like gods. But in fact we are, among other things, animals—emergent parts of the natural order. If our politicians were realists, they would think rather less about missiles and the problem of landing a couple of astronauts on the moon, rather more about hunger and moral squalor and the problem of enabling three billion men, women, and children, who will soon be six billions, to lead a tolerably human existence without, in the process, ruining and befouling their planetary environment.

Animals have no souls; therefore, according to the most authoritative Christian theologians, they may be treated as though they were things. The truth, as we are now beginning to realize, is that even things ought not to be treated as *mere* things. They should be treated as though they were parts of a vast living organism. "Do as you would be done by." The Golden Rule applies to our dealings with nature no less than to our dealings with our fellow-men. If we hope to be well treated by nature, we must stop talking about "mere things" and start treating our planet with intelligence and consideration.

Power politics in the context of nationalism raises problems that, except by war, are practically insoluble. The problems of ecology, on the other hand, admit of a rational solution and can be tackled without the arousal of those violent passions always associated with dogmatic ideology and nationalistic idolatry. There may be arguments about the best way of raising wheat in a cold climate or of re-afforesting a denuded mountain. But such arguments never lead to organized slaughter. Organized slaughter is the result of arguments about such questions as the following: Which is the best nation? The best religion? The best political theory? The best form of government? Why are other people so stupid and wicked? Why can't they see how good and intelligent *we* are? Why do they resist our beneficent efforts to bring them under our control and make them like ourselves?

To questions of this kind the final answer has always been war. "War," said Clausewitz, "is not merely a political act, but also a political instrument, a continuation of political relationships, a carrying out of the same by other means." This was true enough in the eighteen thirties, when Clausewitz published his famous treatise; and it continued to be true until 1945. Now, pretty obviously, nuclear weapons, long-range rockets, nerve gases, bacterial aerosols, and the "Laser" (that highly promising, latest addition to the world's military arsenals) have given the lie to Clausewitz. All-out war with modern weapons is no longer a continuation of previous policy; it is a complete and irreversible break with previous policy.

Power politics, nationalism, and dogmatic ideology are luxuries that the human race can no longer afford. Nor, as a species, can we afford the luxury of ignoring man's ecological situation. By shifting our attention from the now completely irrelevant and anachronistic politics of nationalism and military power to the problems of the human species and the still inchoate politics of human ecology we shall be killing two birds with one stone—reducing the threat of sudden destruction by scientific war and at the same time reducing the threat of more gradual biological disaster.

The beginnings of ecological politics are to be found in the special services of the United Nations Organization. UNESCO, the Food and Agriculture Organization, the World Health Organization, the various Technical Aid Services —all these are, partially or completely, concerned with the ecological problems of the human species. In a world where political problems are thought of and worked upon within a frame of reference whose coordinates are nationalism and military power, these ecology-oriented organizations are regarded as peripheral. If the problems of humanity could be thought about and acted upon within a frame of reference that has survival for the species, the well-being of individuals, and the actualization of man's desirable potentialities as its coordinates, these peripheral organizations would become central. The subordinate politics of survival, happiness, and personal fulfillment would take the place now occupied by the politics of power, ideology, nationalistic idolatry, and unrelieved misery.

In the process of reaching this kind of politics we shall find, no doubt, that we have done something, in President Wilson's prematurely optimistic words, "to make the world safe for democracy."

4.

The Population Bomb

Paul Ehrlich

Professor Paul Ehrlich, one of America's foremost biologists, is a front-line fighter for population control. He is Professor of Biology and Director of the Graduate School of Biological Sciences, Stanford University. His specialty is population biology. He has written over seventy scientific papers and several books on this and related subjects.

Dr. Ehrlich's most recent book, *The Population Bomb*, is a classic of its kind. The author pulls no punches, while his style is vigorous and down to earth.

This book offers its readers the chance to see where they stand vis-à-vis some of the frightening truths of our "brave new world."

What Needs to Be Done

A general answer to the question, "What needs to be done?" is simple. We must rapidly bring the world population under control, reducing the growth rate to zero or making it go negative. Conscious regulation of human numbers must be achieved. Simultaneously we must, at least temporarily, greatly increase our food production. This agricultural program should be carefully monitored to minimize deleterious effects on the environment and should include an effective program of ecosystem restoration. As these projects are carried out, an international policy research program must be initiated to set optimum population-environment goals for the world and to devise methods for reaching these goals. So the answer to the question is simple. Getting the job done, unfortunately, is going to be complex beyond belief—if indeed it can be done. What follows in this chapter are some ideas on how these goals *might* be reached and a brief assessment of our chances of reaching them.

Reprinted from *The Population Bomb*, copyright 1968 by Paul R. Ehrlich. Published by Ballantine Books, Inc.

Getting Our House in Order

The key to the whole business, in my opinion, is held by the United States. We are the most influential superpower; we are the richest nation in the world. At the same time we are also just one country on an ever-shrinking planet. It is obvious that we cannot exist unaffected by the fate of our fellows on the other end of the good ship Earth. If their end of the ship sinks, we shall at the very least have to put up with the spectacle of their drowning and listen to their screams. Communications satellites guarantee that we will be treated to the sights and sounds of mass starvation on the evening news, just as we now can see Viet Cong corpses being disposed of in living color and listen to the groans of our own wounded. We're unlikely, however, to get off with just our appetites spoiled and our consciences disturbed. We are going to be sitting on top of the only food surpluses available for distribution, and those surpluses will not be large. In addition, it is not unreasonable to expect our level of affluence to continue to increase over the next few years as the situation in the rest of the world grows ever more desperate. Can we guess what effect this growing disparity will have on our "shipmates" in the UDCs [underdeveloped countries]? Will they starve gracefully, without rocking the boat? Or will they attempt to overwhelm us in order to get what they consider to be their fair share?

We, of course, cannot remain affluent and isolated. At the moment the United States uses well over half of all the raw materials consumed each year. Think of it. Less than 1/15th of the population of the world requires more than all the rest to maintain its inflated position. If present trends continue, in 20 years we will be much less than 1/15th of the population, and yet we may use some 80% of the resources consumed. Our affluence depends heavily on many different kinds of imports: ferroalloys (metals used to make various kinds of steel), tin, bauxite (aluminum ore), rubber, and so forth. Will other countries, many of them in the grip of starvation and anarchy, still happily supply these materials to a nation that cannot give them food? Even the technological optimists don't think we can free ourselves of the need for imports in the near future, so we're going to be up against it. But, then, at least our balance of payments should improve!

So, beside our own serious population problem at home, we are intimately involved in the world crisis. We are involved through our import-export situation. We are involved because of the possibilities of global ecological catastrophe, of global pestilence, and of global thermonuclear war. Also, we are involved because of the humanitarian feelings of most Americans.

We are going to face extremely difficult but unavoidable decisions. By how much, and at what environmental risk, should we increase our food production in an attempt to feed the starving? How much should we reduce the grain-finishing of beef in order to have more food for export? How will we react when asked to balance the lives of a million Latin Americans against, say, a 30 cent per pound rise in the average price of beef? Will we be willing to slaughter our dogs and cats in order to divert pet food protein to the starving masses in Asia? If these choices are presented one at a time, out of context, I predict that our behavior will be "selfish." Men do not seem to be able to focus emotionally on distant or long-term events. Immediacy seems to be necessary to elicit "selfless" responses. Few Americans could sit in the same room with a child and watch it starve to death. But the death of several million children this year from starvation is a distant, impersonal, hard-to-grasp event. You will note that I put quotes around "selfish" and "selfless." The words describe the behavior only out of context. The "selfless" actions necessary to aid the rest of the world and stabilize the population are our only hope for survival. The "selfish" ones work only toward our destruction. Ways must be found to bring home to all the American people the reality of the threat to their way of life—indeed to their very lives.

Obviously our first step must be to immediately establish and advertise drastic policies designed to bring our own population size under control. We must define a goal of a stable optimum population size for the United States and display our determination to move rapidly toward that goal. Such a move does two things at once. It improves our chances of obtaining the kind of country and society we all want, and it sets an example for the world. The second step is very important, as we also are going to have to adopt some very tough foreign policy positions relative to population control, and we must do it from a psychologically strong position. We will want to disarm one group of opponents at the outset: those who claim that we wish others to stop breeding while we go merrily ahead. We want our propaganda based on "do as we do"—not "do as we say."

So the first task is population control at home. How do we go about it? Many of my colleagues feel that some sort of compulsory birth regulation would be necessary to achieve such control. One plan often mentioned involves the addition of temporary sterilants to water supplies or staple food. Doses of the antidote would be carefully rationed by the government to produce the desired population size. Those of you who are appalled at such a suggestion can rest easy. The option isn't even open to us, thanks to the criminal inadequacy of biomedical research in this area. If the choice now is either such additives or catastrophe, we shall have catastrophe. It might be possible to develop such population control tools, although the task would not be simple. Either the additive would have to operate equally well and with minimum side effects against both sexes, or some way would have to be found to direct it only to one sex and shield the other. Feeding potent male hormones to the whole population might sterilize and defeminize the women, while the upset in the male population and society as a whole can be well imagined. In addition, care would have to be taken to see to it that the sterilizing substance did not reach livestock, either through water or garbage.

Technical problems aside, I suspect you'll agree with me that society would probably dissolve before sterilants were added to the water supply by the government. Just consider the fluoridation controversy! Some other way will have to be found. Perhaps the most workable system would be to reverse the government's present system of encouraging reproduction and replace it with a series of financial rewards and penalties designed to discourage reproduction. For instance, we could reverse our present system of tax exemptions. Since taxes in essence purchase services from the government and since large families require more services, why not make them pay for them? The present system was designed at a time when larger population size was not viewed as undesirable. But no sane society wants to promote larger population size today. The new system would be quite simple (but, of course, not retroactive!). For each of the first two children, an additional $600 would be added to the "taxable income" figure from which the taxes are calculated. For each subsequent child, $1,200 would be added. In order to prevent hardship, minimum levels would be established guaranteeing each family enough for food, clothing, and shelter. Therefore a family with three children and only $4,000 income might pay little or no taxes, but parents making $25,000 who had ten children would pay for their reproductive irresponsibility by forking over the taxes on $35,800. In short, the plush life would be difficult to attain for those with large families—which is as it should be, since they are getting their pleasure from their children, who are being supported in part by more responsible members of society.

On top of the income tax reversal, luxury taxes should be placed on layettes, cribs, diapers, diaper services, expensive toys, always with the proviso that the essentials be available without penalty to the poor (just as free food now is). There would, of course, have to be considerable experimenting on the level of financial pressure necessary to achieve the population goals. To the penalties could be added some incentives. A governmental "first marriage grant" could be awarded each couple in which the age of both partners was 25 or more. "Responsibility prizes" could be given to each couple for each five years of childless marriage, or to each man who accepted irreversible sterilization (vasectomy) before having more than two children. Or special lotteries might be held—tickets going only to the childless. Adoptions could be subsidized and made a simple procedure. Considering the savings in school buildings, pollution control, unemployment compensation, and the like, these grants would be a money-making proposition. But even if they weren't, the price would be a small one to pay for saving our nation.

Obviously, such measures would need coordination by a powerful governmental agency. A federal Department of Population and Environment (DPE) should be set up with the power to take whatever steps are necessary to establish a reasonable population size in the United States and to put an end to the steady deterioration of our environment. The DPE would be given ample funds to support research in the areas of population control and environmental quality. In the first area it would promote intensive investigation of new techniques of birth control, possibly leading to the development of mass sterilizing agents such as were discussed above. This research will not only give us better methods to use at home; they are absolutely essential if we are to help the UDCs to control their populations. Many peoples lack the incentive to use the Pill. A program requiring daily attention just will not work. This is one reason why a Papal decision to accept the Pill but not other methods of birth control would be only a small step in the right direction. The DPE also would encourage more research on human sex determination, for if a simple method could be found to guarantee that first-born children were males, then population control problems in many areas would be somewhat eased. In our country and elsewhere couples with only female children "keep trying" in hope of a son.

Two other functions of the DPE would be to aid Congress in developing legislation relating to population and environment, and to inform the public of the needs for such legislation. Some of these needs are already apparent. We need a federal law guaranteeing the right of *any* woman to have an abortion if it is approved by a physician. We need federal legislation guaranteeing the right to voluntary sterilization for both sexes and protecting physicians who perform such operations from legal harassment. We need a federal law requiring sex education in schools—sex education that includes discussion of the need for regulating the birth rate and of the techniques of birth control. Such education should begin at the earliest age recommended by those with professional competence in this area—certainly before junior high school.

By "sex education" I do not mean courses focusing on hygiene or presenting a simple-minded "birds and bees" approach to human sexuality. The reproductive function of sex must be shown as just one of its functions, and one that must be carefully regulated in relation to the needs of the individual and society. Much emphasis must be placed on sex as an interpersonal relationship, as an important and extremely pleasurable aspect of being human, as mankind's major and most enduring recreation, as a fountainhead of his humor, as a phenomenon that affects every aspect of his being. Contrary to popular mythology, sex is one of man's *least* "animal" functions. First of all, many animals (and plants) get along without any sex whatsoever. They reproduce asexually. It is clear from biological research that sex is not primarily a mechanism of reproduction; it is a mechanism that promotes variability. In many organisms which do have sexual processes, these processes occur at a stage in the life cycle that is not the stage at which reproduction occurs. And, of course, no other animal has all of the vast cultural ramifications of sex that have developed in human society. In short, sex, as we know it, *is a peculiarly human activity.* It has many complex functions other than the production of offspring. It is now imperative that we restrict the reproductive function of sex while producing a minimum of disruption in the others.

Fortunately, there are hopeful signs that the anti-human notions that have long kept Western society in a state of sexual repression no longer hold sway over many of our citizens. With a rational atmosphere mankind should be able to work out the problems of de-emphasizing the reproductive role of sex. These problems include finding substitutes for the sexual satisfaction which many women derive from child-bearing and finding substitutes for the ego satisfaction that often accompanies excessive fatherhood. A rational atmosphere should also make it easier to deal with the problems of venereal disease and of illegitimacy. The role of marriage would become one of providing the proper environment for the rearing of wanted children. All too often today marriage either provides a "license" for sexual activity or a way of legitimizing the results of premarital sexual activity.

If we take the proper steps in education, legislation, and research, we should be able in a generation to have a population thoroughly enjoying its sexual activity, while raising smaller numbers of physically and mentally healthier children. The population should be relatively free of the horrors created today by divorce, illegal abortion, venereal disease, and the psychological pressures of a sexually repressive and repressed society. Much, of course, needs to be done, but support for action in these directions is becoming more and more common in the medical profession, the clergy, and the public at large. If present trends can be continued, we should be able to minimize and in some cases reverse social pressures against population control at home and to influence those abroad in the same direction.

Of course, this enlightened atmosphere does not exist today. Potent forces still must be overcome if we are to get the attitude of our government changed in the area of population control. Although the performance and attitudes of American Catholics relative to the use of birth control are similar to those of non-Catholics, conservative elements in the Church hierarchy still resist change. The degree to which this resistance goes against the attitudes of American Catholics was revealed in a Gallup Poll taken in late 1965. Of the Catholics questioned, 56 percent expressed the opinion that the Church should change its opinion on methods of birth control, while only 33 percent thought it should not change. Opinion among intellectual Catholics seems even more heavily in favor of a change in the Church's position.

Recently Pope Paul VI formed a commission to study in detail the question of birth control. The majority opinion of this commission was that the practice of birth control by means other than those involving abortion was wholly consistent with the Catholic view of marriage and sexuality, and that the method of birth control to be used can best be decided by the couple concerned. This opinion, the commission felt, fit directly into the teaching on responsible parenthood of the Second Vatican Council.

Since this report should have dissolved any fears the Pope may have had about contraception, it is a mystery to informed Catholics why he has not acted.

A Catholic colleague, Dr. John H. Thomas, recently wrote to me:

"My first duty as a Catholic is to do what I believe is morally correct. There is no doubt in my mind that the position of the Church with respect to birth control is morally wrong. The price of doctrinaire insistence on unworkable methods of birth control is high. It contributes to misery and starvation for billions, and perhaps the end of civilization as we know it. As a scientist I also know that Catholic doctrine in this area is without biological foundation. It is therefore my duty both to myself and to the Church not just to ignore this doctrine, but to do everything within my power to change it. After all, without drastic worldwide measures for population control in the near future, there will be no Church anyway. If the Church, or for that matter, any organized religion, is to survive, it must become much more humanitarian in focus. If it does, the theology will take care of itself."

All these are hopeful signs, but unhappily the Church hierarchy and certain conservative Catholic groups are still hard at work to raise the death rate by fighting *effective* moves to lower the birth rate. It takes a great deal of patience for a biologist familiar with the miseries of overpopulation to read through documents that represent the views of even "enlightened" Catholics. For instance, consider the views of Dr. Donald N. Barrett, Professor of Sociology at Notre Dame. Married and the father of ten children, Dr. Barrett testified before Senator Gruening's Subcommittee on March 2, 1966. The testimony was given on a bill to coordinate and disseminate birth control information upon request. Dr. Barrett makes the strong point that action in population control must avoid government coercion (he makes the distinction that the Catholic Church in this matter is only "morally coercive"). He is unwilling to cooperate with any group which supports abortion or sterilization as birth control methods. He puts much weight on refinements of the rhythm method, which is patently hopeless as a population control method. Much emphasis is put on "licit" methods. The "moral coercion" of the Church is present in veiled threats to the Subcommittee not to take positions that would prevent a consensus acceptable to the Catholic Church. Barrett's testimony was notably free of comment on the physical coerciveness of starvation, plague, and thermonuclear war. Nor does he comment on the human values that will be lost when civilization goes down the drain. I wonder what Barrett thinks of his Church's role in keeping help, in the form of contraceptive education and devices, from those miserable people in Colombia. How hollow his talk of "moral coercion" would sound down there! You might note that if Barrett's descendants continued his rate of propagation for just ten generations, they would number in the tenth generation ten billion people—three times the entire population of the Earth today.

The testimony of William H. Ball, general counsel for the Pennsylvania Catholic Conference, before Senator Gruening's committee also makes interesting, if depressing, reading. Its tenor can be guessed by a quote from Ball's article in *Commonweal* entitled "The Court and Birth Control." He states in that article that the government "should be neutral, neither penalizing birth control nor promoting it." In his testimony he expresses great concern for "freedom from governmental inquisition, the related right of privacy, concern for the weaker members of society," and "governmental coercion of mind and conscience." Curiously, though, he does not discuss the great increase in governmental interference and restriction of freedom that has accompanied population growth. He does not mention what has been happening to privacy as population increases. He does not talk about protecting the poor of the world from starvation that is a direct result of population pressure. Let me quote the phrases from his testimony that most thoroughly attest to the total absurdity of his view:

"The repetition of the term 'population' throughout the text, the reference in the bill to population 'control,' *coupled with an absence of any indication of means other than population control as a solution to problems of population growth. . . .* " (My emphasis). It will hardly come as a shock to you that he says a little further on, "At the time when the Congress contemplates embarking the nation upon so unprecedented a program, the Pennsylvania Catholic Conference feels it its duty to state its conviction that the public power and public funds should not be used for the providing of birth control services."

In short, Catholic witnesses are opposed even to attempts to institute inadequate federal programs of population control. Catholic politicians at home and abroad operate in many ways to obstruct population control. They often effectively block action on population control at the international level. And population control, of course, is the *only* solution to problems of population growth. Unless the Pope does a complete about-face, I think we can count on continuing an effective Catholic support for raising the death rate.

This encouragement of high death rates through political interference is now the most important role of the Church in the population crisis. There is little reason to believe that, if obstructionist behavior by the hierarchy and other influential Catholics ceased, *performance* of Catholic couples would differ significantly from that of non-Catholics in most areas. Furthermore, in the UDCs outside of Latin America Catholics are rarely a significant portion of the problem. It is a mistake to focus too strongly on the Catholic element in the population situation. True, we must bring pressure to bear on the Pope in hope of getting a reversal of the Church's position. Probably the best way is to support those American Catholics who already realize that opposition to birth control is automatically support for increased misery and death. If such a reversal can be obtained, mankind's chances for survival will improve somewhat, and millions upon millions of Catholics will be able to lead better lives. But the population problem will not be "solved."

Biologists must promote understanding of the facts of reproductive biology which relate to matters of abortion and contraception. They must do more than simply reiterate the facts of population dynamics. They must point out the biological absurdity of equating a zygote (the cell created by joining of sperm and egg) or fetus (unborn child) with a human being. As Professor Garret Hardin of the University of California pointed out, that is like confusing a set of blueprints with a building. People are people because of the interaction of genetic information (stored in a chemical language) with an environment. Clearly, the most "humanizing" element of that environment is the cultural element, to which the child is not exposed until after birth. When conception is prevented or a fetus destroyed, the *potential* for another human being is lost, but that is all. That potential is lost *regardless* of the reason that conception does not occur—there is no biological difference if the egg is not fertilized because of timing, or because of mechanical or other interference.

Biologists must point out that contraception is for many reasons more desirable than abortion. But they must also point out that in many cases abortion is much more desirable than childbirth. Above all, biologists must take the side of the hungry billions of *living* human beings today and tomorrow, not the side of *potential* human beings. Remember, unless numbers are limited, if those potential human beings are born, they will at best lead miserable lives and die young. We cannot permit the destruction of humanity to be abetted by a doctrine conceived in total ignorance of the biological facts of life.

Basically, I think the Catholic situation is much more amenable to solution than that associated with our current views of economics. The winds of change are clearly blowing in religion—blowing too late, perhaps, but blowing. Yet the idea of an over-expanding economy fueled by population growth seems tightly entrenched in the minds of businessmen, if not in the minds of economists. Each new baby is viewed as a consumer to stimulate an ever-growing economy. Each baby is, of course, potentially one of the unemployed, but a consumer nonetheless. The Rienows estimate that each American baby will consume in a 70-year life span, directly or indirectly: 26 million gallons of water, 21 thousand gallons of gasoline, 10 thousand pounds of meat, 28 thousand pounds of milk and cream, $5,000 to $8,000 in school building materials, $6,300 worth of clothing, and $7,000 worth of furniture. It's not a baby, it's Superconsumer!

Our entire economy is geared to growing population and monumental waste. Buy land and hold it; the price is sure to go up. Why? Exploding population on a finite planet. Buy natural resources stocks; their price is sure to go up. Why? Exploding population and finite resources. Buy automotive or airline stocks; their price is sure to go up. Why? More people to move around. Buy baby food stocks; their price is sure to go up. Why? You guess. And so it goes. Up goes the population and up goes that magical figure, the Gross National Product (GNP). And, as anyone who takes a close look at the glut, waste, pollution, and ugliness of America today can testify, it is well-named—as *gross* a product as one could wish for. We have assumed the role of the robber barons of all time. We have decided that we are the chosen people to steal all we can get of our planet's gradually stored and limited resources. To hell with future generations, and to hell with our fellow human beings today! We'll fly high now—hopefully they'll pay later.

We thought the game would end only in hundreds or even thousands of years through resource depletion. But the bill is coming due before we expected it. Now we find that to be among the least of our problems. The poor of the world show signs of not being happy with our position. Indeed even the poor at home seem a little ill-disposed toward our behavior. Maybe we can hold them down by force, you say. Maybe so—that remains to be seen. It is likely to be uneconomical to do so. Besides, what has been properly called "the *effluent* society" shows signs of strangling itself without the intervention of enraged "have-nots." Will our gross national product soon be reduced to no national product?

The answer is that it surely will unless we take a hard look at our present economic system. There are some very distinguished economists who do not feel that our capitalist system must be fueled by an ever-growing population or even-continuing depletion of resources (both of which are impossible, anyway). There, in fact, seems to be no reason why the GNP cannot be kept growing for a very long time *without population growth*.

Dr. J. J. Spengler recently wrote:

"In the future, economic growth will depend mainly upon invention, innovation, technical progress, and capital formation, upon institutionalized growth-favoring arrangements. Population growth will probably play an even smaller role than I have assigned it in earlier discussion. *It is high time, therefore, that business cease looking upon the stork as a bird of good omen.*" (My emphasis.)

Ways must be found to promote the idea that problems associated with population growth will more than cancel the "advantages" of financial prosperity. Perhaps the best way to do this would be to encourage Americans to ask exactly what our financial prosperity is for. What will be done with leisure time and money when all vacation spots are crowded beyond belief? Is it worth living in the Los Angeles smog for 50 weeks in order to spend two weeks in Yosemite Valley—when the Valley in the summer may be even more crowded than L.A. and twice as smoggy? What good is having the money for a fishing trip when fish are dead or poisonous because of pesticide pollution? Why own a fancy car in which to get asphyxiated in monster traffic jams? Do we want more and more of the same until we have destroyed ourselves? Sizable segments of our population, especially the young, are already answering that question: "Hell, no!" Their response should be considered carefully by population-promoting tycoons.

Obviously, the problem of our deteriorating environment is tied in very closely with the overall economic problem. We must reverse the attitudes so beautifully exemplified by one of our giants of industry when he said that "the ability of a river to absorb sewage is one of our great natural resources and should be utilized to the utmost." Legal steps must be taken, and taken fast, to see to it that polluters pay through the nose for their destructive acts. The old idea that industry could create the mess and then the taxpayers must clean it up has to go. The garbage produced by an industry is the responsibility of that industry. The government should not use other people's money to clean it up. Keep the government out of business. Let it play its proper role in a capitalistic society—seeing to it that all segments of private enterprise do business honestly, seeing to it that the interests of the fishing industry are not subordinated to those of the petrochemical industry, seeing to it that your right to swim in a public lake is not subordinated to the desire of a steel company to make an inflated profit.

The policeman against environmental deterioration must be the powerful Department of Population and Environment mentioned above. It must be carefully insulated against the forces that will quickly be aligned against it. It is going to cost industry money. It is going to cost municipalities money. It is going to hit a lot of us where it hurts. We may have to do without two gas-gulping monster cars per family. We may have to learn to get along with some insect damage in our produce. We may have to get along with much less fancy packaging of the goods we purchase. We may have to use cleaners that get our clothes something less than "whiter than white." We may have to be satisfied with slower coast-to-coast transportation. Such may be the cost of survival. Of course, we may also have to get along with less emphysema, less cancer, less heart disease, less noise, less filth, less crowding, less need to work long hours or "moonlight," less robbery, less assault, less murder, and less threat of war. The pace of life may slow down. We may have more fishing, more relaxing, more time to watch TV, more time to drink beer (served in bottles that *must* be returned).

The Department of Population and Environment (DPE) would place extremely strict controls on the use of dangerous pesticides and would encourage research on economically more reasonable methods of control. We have barely scratched the surface in what can be done with biological controls, including ways of manipulating the genetics of populations. We do not know enough about the ways the chemical and biological controls might be integrated in ecologically intelligent ways. But perhaps the greatest service the DPE might perform immediately from its inception would be to expose the stupidity and futility of today's pesticide practices. Any properly constituted DPE would have a strong complement of systems ecologists—ecologists who use the methods of operations research and systems analysis to evalute complex ecological systems. As a foretaste of what the DPE might say, let me quote to you from a letter I recently received from Professor K. E. F. Watt of the University of California, one of today's outstanding systems ecologists:

"... most control programs are set up without a threshold; that is, spray is used each season whether significant densities of pests are present or not. Thus, this is an example of a business providing the amazing spectacle of supporting an overhead which is not associated with a corresponding marginal increase in gross profit. It is this type of practice which has led many fruit orchard owners into such dire economic straits that they have had to sell their land for housing projects or factory sites.

"It is most important to point out to the public that a pest control program should have two consequences: (1) either plant or animal being attacked by the pests should be saved, and (2) there should be fewer pests in subsequent generations following treatment. The yardstick by which all control programs should be evaluated comes from those dramatically successful programs in which plants or animals were saved, and pests declined in density. You are aware of examples that provide this yardstick . . . the Florida screwworm study is a prize example. *By this criterion most pesticide projects have been failures.*" (My emphasis.)

The DPE would also be responsible for pushing legislation to stop the wasting of resources. It would move toward creating a vast waste recovery industry, an industry that might well make "trash" obsolete. Reusable containers might be required by law for virtually all products, as was recently suggested by Dr. Athelstan Spilhaus. He points out the necessity of controlling trash and pollutants at the source, stating, "Regardless of what any economist tells me, I'm convinced by the second law of thermodynamics that it must be cheaper to collect something at the source than to scrape it off the buildings, wash it out of the clothes, and so forth." There's that old, immutable law again. If the product is deteriorated and scattered, usable energy has been lost, and more must be injected into the system if order is to be restored, by either collecting or reconstituting the product. The less deterioration or scattering we permit, the less energy we must use. And energy is expensive.

The DPE would have to take a good hard look at our energy budget, especially at the rate at which we are expending our irreplaceable fossil fuels. It would have to evaluate carefully the possible role of atomic fission or fusion in replacing fossil fuels as an energy source. It would have to evaluate hydroelectric power in relation to the other two. These sources cannot be considered in isolation. Atomic facilities must have their waste disposal problems integrated into the evaluation. Hydroelectric power must be considered in a framework of the gradual altering of the ecology of rivers and flood plains and of Earth's topography through the building and silting up of dams. It must be considered in relationship to salmon fisheries and downstream farming. Both atomic and hydroelectric power must be considered in relation to the expenditures of fossil fuels required to mine, transport, and process the metals and concrete from which facilities are built. That we are presently living beyond our means is obvious from the simple fact that we are madly depleting nonreplenishable resources. Careful plans must be laid for getting the Earth back in balance, on the hopeful assumption that some way can be found to avoid the doom now confronting us.

By now you are probably fed up with this discussion. Americans will do none of these things, you say. Well, I'm inclined to agree. As an eternal optimist, however, I will provide some suggestions in the last chapter of this book for what you might do to improve the chances that action will be taken. Improve them from, say, one in a thousand to one in a hundred, but improve them. Meanwhile let's make the unlikely assumption that this country will turn aside from its suicidal course and start a sensible domestic program of population and environmental control. How can we then help with the world problem?

Realism and International Aid

Once the United States has adopted sane policies at home, we will be in a position to take the lead in finding a solution to the problem on a world scale. What we will need first and foremost is a plan that will produce a maximum amelioration of the time of famines with the relatively limited resources we have in hand. Even drastic population control measures need decades to work, and we do not have the capacity to feed the needy of the world over the next decade or so. Our giant food surpluses are gone, and even at maximum production we would not be able to produce surplus enough for all (to say nothing of getting it properly distributed). In addition, we are the only country which will be in a position to give away food. Canada, Australia, Argentina, and the other few countries with exportable surpluses will be largely occupied with selling food to hungry countries that are in a position to pay. These granary countries will need the income that they earn in this way, or the goods they can receive in exchange for food. The UDCs cannot expect major charity from them.

What kind of policies should we be designing to guide our actions during the time of famines? To my knowledge, there has been only one realistic suggestion in this area—a policy proposed by William and Paul Paddock in their book *Famine—1975!* The Paddocks suggest an American policy based on the concept of "triage" borrowed from military medicine. The idea briefly is this: When casualties crowd a dressing station to the point where all cannot be cared for by the limited medical staff, some decisions must be made on who will be treated. For this purpose the triage system of classifcation was developed. All incoming casualties are placed in one of three classes. In the first class are those who will die regardless of treatment. In the second class are those who will survive regardless of treatment. The third contains those who can be saved only if they are given prompt treatment. When medical aid is limited, it is concentrated only on the third group—the others are left untreated.

The Paddocks suggest that we devise a similar system for classifying nations. Some will undergo the transition to self-sufficiency without drastic aid from us. They will be ones with abundant money for foreign purchases, or with efficient governments, strong population control programs, and strong agricultural development programs. Although our aid might help them, they could get along without it. The Paddocks suggest that Libya is probably such a country. It has the resources, in the form of oil, that will allow it to purchase food as its population booms. If analysis shows them to be correct, we should withhold food aid from Libya.

Some nations, on the other hand, may become self-sufficient if we give them help. They have a chance to make it if we can give them some food to tide them over. The Paddocks think that Pakistan, at least West Pakistan, may be such a country. Others to whom I have spoken agree. Our food aid may give the Pakistani government, under the tough-minded leadership of President Ayub Khan, time to press home its population control and agricultural development programs. If they are right, we should continue to ship food to Pakistan.

Finally there is the last tragic category—those countries that are so far behind in the population-food game that there is no hope that our food aid will see them through to self-sufficiency. The Paddocks say that India is probably in this category. If it is, then under the triage system she should receive no more food.

The Paddocks' views have not, to say the least, been greeted with enthusiasm by the Indian govenment. Nor have their views been applauded by those in our government whose jobs depend on the willy-nilly spreading of American largess abroad, or by the assorted do-gooders who are deeply involved in the apparatus of international food charity. India, as we noted earlier, blames its current problems on bad monsoons (which indeed did occur). It has conveniently forgotten that the Indian govenment itself predicted in 1959 that a serious gap would appear between food production and population in 1965–1966. At any rate, the Indian government seems deeply concerned about the possibility that the Paddocks' idea might take hold and that India will be denied further food.

In my opinion, there is no rational choice *except* to adopt some form of the Paddocks' strategy as far as food distribution is concerned. I have incorporated a version of it in a broader plan I am suggesting below. The Paddocks deserve immense credit for their courage and foresight in publishing *Famine—1975!*, which may be remembered as one of the most important books of our age. They will receive criticism from certain segments of our society. They will offend the groups which discounted the warnings of a decade or more ago, warnings that we would be in serious trouble today unless the population was brought under control. Criticism from those groups is a compliment.

What might be a possible strategy leading to man's passage with minimum casualties through the time of famines? Obviously, if we are to find a long-range solution, the full weight of the resources of the United States and the other DCs must be brought to bear. My suggestion would be that the United States, Russia, Great Britain, Canada, Japan, Australia, Europe, and other DCs immediately set up, through the United Nations, a machinery for "area rehabilitation." This plan would involve simultaneous population control, agricultural development, and, where resources warrant it, industrialization of selected countries or sections of countries. The bedrock requirement of this program would be population control, necessarily including migration control to prevent swamping of aided areas by the less fortunate. Of course, the size of the areas covered would be dependent in no small part on the scale and effectiveness of the effort made by the developed countries. Hopefully, we can persuade the United States to lead the way. So far our efforts toward aiding the UDCs have, in terms of the percentage of our gross national product committed, been behind that of many other DCs, who can less well afford it.

The specific requirements of the program would vary from area to area. Possibly the first step in all areas would be to set up relay stations and distribute small transistorized TV sets to villages for communal viewing of satellite-transmitted programs. We must have channels for reaching the largely rural populations of the "other world." TV programs would explain the rehabilitation plan for each area. These programs would have to be produced with the combined skill of Madison Avenue, of people with great expertise in the subjects to be presented, and of people with intimate knowledge of the target population. The programs could be presented both "straight" and as cleverly devised "entertainment." They would introduce the UDC populations to the need for agricultural innovations and explain public health measures. The programs would use the prospect of increased affluence as a major incentive for gaining cooperation. It seems unlikely that the threat of future starvation would have much impact. If necessary, however, the TV channel could be used to make it clear that the continuance of food supplies depends on the cooperation of the people in the area. Perhaps they could be made to realize that only by making progress toward population control and self-sufficiency can they avoid disaster.

Other steps would vary a great deal from place to place. In some agricultural areas needs would be well enough known for assistance to start immediately, perhaps with "on site" training of agricultural technicians. Such a program could lead to a sort of "county agricultural agent" system in which trained people work closely with farmers. These systems have proved their great worth in many parts of the world. Schools to train agents and other agricultural personnel, including farmers brought in on rotation, would be of immense value to the agriculture of most UDCs.

In some places the problems of agriculture would be so severe that research stations, manned by teams from the DCs, might have to function for a decade or so before local agriculture had a chance of being revolutionized. "Improved" strains of various crops developed elsewhere might not grow satisfactorily or might be unacceptable to the local people as food. Since the supply of trained people in DCs suitable for running stations doing research in exotic agriculture is limited, priority systems for station establishment must be set up. At the same time ways must be found to increase the supply of agricultural scientists being trained in the DCs.

In all areas studies should be initiated to determine how much agricultural and industrial development is feasible. Demographers must determine how many people, at each stage of development, can live reasonably comfortable, secure lives. That is, demographic goals must be set that are reasonable in the light of each country's basic resources. Unless demographic goals are set and met, the entire program will inevitably fail. Population control must be made to work, or all our other efforts will have been in vain.

Needless to say, the sociopolitical problems of initiating such a program would be colossal. It might not, for instance, be feasible to operate through the United Nations, because countries will not all be aided equally. This problem might be sidestepped by using the "area" concept rather than strictly political units. Thus, if migration could be controlled, some sections of India might be aided and others not. Perhaps we should support secessionist movements in UDCs when the group departing is better developed than the previous political unit as a whole. Perhaps we should have supported Katanga, not the Congo. Perhaps we should now support Biafra, not Nigeria. West Pakistan might receive aid, but not East Pakistan. It might be to our advantage to have some UDCs more divided or even rearranged, especially along economic axes. After all, most political boundaries in Southern Asia and Africa reflect not economically viable units, but the conflicting interests of European powers 75 years ago. I know this all sounds very callous, but remember the alternative. The callous acts have long since been committed by those who over the years have obstructed a birth rate solution or downgraded or ignored the entire problem. Now the time has come to pay the piper, and the same kind of obstructionists remain. If they succeed, we will all go down the drain.

While we are working toward setting up a world program of the general sort outlined above, the United States could take effective unilateral action in many cases. A good example of how we might have acted can be built around the Chandrasekhar incident I mentioned earlier. When he suggested sterilizing all Indian males with three or more children, we should have applied pressure on the Indian government to go ahead with the plan. We should have volunteered logistic support in the form of helicopters, vehicles, and surgical instruments. We should have sent doctors to aid in the program by setting up centers for training para-medical personnel to do vasectomies. Coercion? Perhaps, but coercion in a good cause. I am sometimes astounded at the attitudes of Americans who are horrified at the prospect of our government insisting on population control as the price of food aid. All too often the very same people are fully in support of applying military force against those who disagree with our form of government or our foreign policy. We must be relentless in pushing for population control around the world.

I wish I could offer you some sugar-coated solutions, but I'm afraid the time for them is long gone. A cancer is an uncontrolled multiplication of cells; the population explosion is an uncontrolled multiplication of people. Treating only the symptoms of cancer may make the victim more comfortable at first, but eventually he dies—often horribly. A similar fate awaits a world with a population explosion if only the symptoms are treated. We must shift our efforts from treatment of the symptoms to the cutting out of the cancer. The operation will demand many apparently brutal and heartless decisions. Then pain may be intense. But the disease is so far advanced that only with radical surgery does the patient have a chance of survival.

So far I have talked primarily about the strategy for easing us through the hazardous times ahead. But what of our ultimate goals? That, of course, is something that needs a great deal of discussion in the United States and elsewhere. Obviously, we need a stable world population with its size rationally controlled by society. But what should the size of that population be? What is the optimum number of human beings that the Earth can support? This is an extremely complex question. It involves value judgments about how crowded we should be. It also includes technical questions of how crowded we *can* be. Research should obviously be initiated in both areas immediately.

If we are to decide how crowded we should be, we must know a great deal more about man's perception of crowding and about how crowding affects human beings. Certainly people in different cultures and subcultures have different views of what densities of people (people per unit area) constitute crowding under different conditions. But what exactly are those densities and conditions? Under what conditions do people consider themselves neither crowded nor lonely? Research on these questions has barely been started. It must be accompanied by studies of how crowding affects people, including both "overcrowding" (too many people per unit area) and "undercrowding" (too few per unit area). These problems are more difficult to study, especially since the effects of crowding are often confounded by poverty, poor diet, unattractive surroundings, and other related phenomena.

But difficult as these problems are, they must be investigated. We know all too well that when rats or other animals are overcrowded, the results are pronounced and usually unpleasant. Social systems may break down, cannibalism may occur, breeding may cease altogether. The results do not bode well for human beings as they get more and more crowded. But extrapolating from the behavior of human beings is much more risky than extrapolating from the physiology of rats to the physiology of human beings. Man's physical characteristics are much more ratlike than are his social systems. This research must be done on man.

Within the limits imposed by nature, I would view an optimum population size for the Earth to be one permitting any individual to be as crowded or as alone as he or she wished. Enough people should be present so that large cities are possible, but people should not be so numerous as to prevent people who so desire from being hermits. Pretty idealistic, but not impossible in theory. Besides, some pretty far-reaching changes are going to be required in human society over the next few deacdes, regardless of whether or not we stop the population explosion. We've already reached a density at which many of our institutions no longer function properly. As the distinguished historian, Walter Prescott Webb, pointed out 16 years ago, with the closing of the World Frontier, a set of basic institutions and attitudes became outdated. When the Western Hemisphere was opened to exploitation by Europeans, a crowded condition suddenly was converted into an uncrowded one. In 1500 the ratio of people to available land in Europe was estimated to have been about 27 people per square mile. The addition of the vast, virtually unpopulated frontier of the New World moved this ratio back down to less than five per square mile. As Webb said, the frontier was, in essence, "a vast body of wealth without proprietors." Europeans moved rapidly to exploit the spatial, mineral, and other material wealth of the New World. They created an unprecedented economic boom that lasted some 400 years. The boom is clearly over, however, at least as far as land is concerned. The man/land ratio went beyond 27 people per square mile again before 1930. Since all of the material things on which the boom depended also come ultimately from the land, the entire boom is also clearly limited. Of course, how to end that boom gracefully, without the most fantastic "bust" of all time, is what this book is all about.

Somehow we've got to change from a growth-oriented, exploitative system to one focused on stability and conservation. Our entire system of orienting to nature must undergo a revolution. And that revolution is going to be extremely difficult to pull off, since the attitudes of Western culture toward nature are deeply rooted in Judeo-Christian tradition. Unlike people in many other cultures, we see man's basic role as that of dominating nature, rather than as living in harmony with it. This entire problem has been elegantly discussed by Professor Lynn White, Jr., in *Science* magazine. He points out, for instance, that before the Christian era trees, springs, hills, streams, and other objects of nature had guardian spirits. These spirits had to be approached and placated before one could safely invade their territory. As White says, "By destroying pagan animism, Christianity made it possible to exploit nature in a mood of indifference to the feelings of natural objects." Christianity fostered the wide spread of basic ideas of "progress" and of time as something linear, nonrepeating, and absolute, were foreign to the Greeks and Romans, who had a cyclical (repeating) view of time and could not envision the world as having a beginning. Although a modern physicist's view of time might be somewhat closer to that of Greeks than the Christians, it is obvious that the Christian view is the one held by most of us. God designed and started the whole business for our benefit. He made a world for us to dominate and exploit. Our European ancestors had long since developed the "proper" attitudes when the opportunity to exploit the New World appeared.

Both science and technology can clearly be seen to have their historical roots in natural theology and the Christian dogma of man's rightful mastery over nature. Therefore, as White claims, it is probably in vain that so many look to science and technology to solve our present ecological crisis. Much more basic changes are needed, perhaps of the type exemplified by the much-despised "hippie" movement—a movement that adopts most of its religious ideas from the non-Christian East. It is a movement wrapped up in Zen Buddhism, physical love, and a disdain for material wealth. It is small wonder that our society is horrified at hippie's behavior—it goes against our most cherished religious and ethical ideas. I think it would be well if those of us who are totally ensnared in the non-hip part of our culture paid a great deal of attention to the movement, rather than condemn it out of hand. They may not have *the* answer, but they may have *an* answer. At the very least they are asking the proper questions. Here is what White, a churchman, has to say: "Both our present science and our present technology are so tinctured with orthodox Christian arrogance toward nature that no solution for our ecologic crisis can be expected from them alone. Since the roots of our trouble are so largely religious, the remedy must also be essentially religious, whether we call it that or not."

So there is considerable reason for believing that extremely fundamental changes in our society are going to be required in order to preserve any semblance of the world we know. Furthermore, those changes are going to have to take place in a framework of certain natural limits. For, as I hope I have convinced you, even though we would like to dominate nature, it still dominates us!

What are those limits that are imposed by nature? We don't know exactly. Finding out will involve complex questions of energy sources and the availability of the materials necessary for the production of food. There is some disagreement as to exactly how dependent upon fossil fuels we shall remain and what the ultimate consequences of their depletion beyond certain levels will be. But at a minimum it seems safe to say that a population of one or even two billion people could be sustained in reasonable comfort for perhaps 1,000 years if resources were husbanded carefully. A mere century of stability should provide ample time to investigate most technological leads and to do the social adjusting and policy planning necessary to set realistic goals on a more or less permanent basis. Our big problem today is how to bring the population under control, reduce its size to that general range, and create the atmosphere in which necessary changes, investigations, and planning can take place. If we are not successful in reducing the population size, but do stabilize it at perhaps four or five billion, we will still have a chance. Of course, mankind's options will be fewer and people's lives almost certainly less pleasant than if the lower figure is attained.

The Chances of Success

Many of you are doubtless saying now, "It's too unrealistic—it can't be done." I think you're probably right—as I said earlier, the chances of success are small. Indeed, they are probably infinitesimal if success is to be measured only by the initiation of a complete program such as I have suggested. But partial programs can help. Indeed, even if the worst happens, short of the end of civilization, efforts toward solving the population problem may not be in vain. Suppose we do not prevent massive famines. Suppose there are widespread plagues. Suppose a billion people perish. At least if we have called enough attention to the problem, we may be able to keep the whole mess from recycling. We must make it impossible for people to blame the calamity on too little food or technological failures or "acts of God." They must at least face the essential cause of the problem—overpopulation.

What Can You Do?

The question I am most frequently asked after giving talks about the population explosion is, "What can I do to help?" The obvious first answer is, "Set an example—don't have more than two children." That reply really sets the pace, because I am becoming more and more convinced that the only real hope in this crisis lies in the grass-roots activities of individuals. We must change public opinion in this country, and through public opinion change the direction of our government. The fact that we cannot count on vast funds to support our efforts does not have to be an unsurmountable obstacle. In the five years that I have been a part-time propagandist, I have found that many people in influential positions share my concern. I have had encouraging letters from all over the world. People in radio and television have been extremely helpful in providing exposure for the issues. Exposure for the issues, however, is not enough. We must create enough pressure to convince politicians that their political survival is at stake unless they get behind some really effective measures to deal with mankind's most pressing problem.

5.

Ecological Upsets: Climate and Erosion

Tom Stonier

Professor Stonier is currently engaged in teaching and biological research at Manhattan College in the Bronx, New York. He has been on the staff of the Rockefeller Institute and has lectured on the effects of nuclear weapons to many scientific, medical, educational, and other groups, including the American Association for the Advancement of Science and the New York and Washington Academies of Science.

In his frightening but realistic book *Nuclear Disaster*, he assesses for the first time the physical, biological, economic and social consequences of such a catastrophy in their interaction with one another. This book forces us to think about the unthinkable.

The ecological imbalances that could be produced by extensive fallout have been described in the preceding chapter. These imbalances would be caused by the radiation emitted by the fallout. In addition, the simple act of injecting a large quantity of dirt and other matter into the atmosphere could also have profound consequences. First, fallout particles settling on snow would cause the snow to melt more rapidly. Under certain circumstances this could lead to disastrous floods. Second, and more important, injecting large quantities of small particles high into the atmosphere could produce marked changes in the weather, changes which under extreme conditions might initiate another ice age.

Let us examine these possibilities in somewhat more detail: Dirt particles on snow greatly enhance solar heat absorption by the snow cover. Farmers in northern and eastern Europe take advantage of this fact by spreading coal dust on snow-covered fields to insure early melt out, thereby prolonging the growing season. Irregular patterns of melting snow may also be observed in western Colorado, caused by dust blown in, presumably from Utah.

Thus, should a substantial amount of fallout settle in the mountains during the winter, and if the ground beneath the snow cover is frozen, then the usual spring flood conditions would be seriously aggravated. Although at first glance this possibility seems somewhat academic, the fact that many communities are located in the valleys of rivers that flood regularly means that great havoc could be wrought by extensive fallout deposition upstream. The material spread over the countryside by a single twenty-megaton surface burst on soft soil could theoretically cover an area of more than 3,600 square miles to a depth of one millimeter.[1] One could

[1] A twenty-megaton surface burst on soil would produce a crater 800 feet deep and 3,400 feet across, displacing a volume of 3.6 x 109 cubic feet. If one assumes that only 10 per cent of the displaced material is deposited over the countryside as small fallout particles (the rest comprising the crater lip, throw out, stem material, and material injected into the atmosphere), then 3.6 x 108 cubic feet would be involved. theoretically, one cubic foot of fine particles could uniformly cover an area of about 300 square feet to a depth of one millimeter. therefore, the material spread over the countryside by a twenty-megaton surface burst

readily envision that if sixty-six megatons were detonated in a broad belt from Pittsburgh to Cleveland, as was assumed by the 1959 Holifield Hearings, but late in the winter, that sizable areas of the snow-covered Appalachian Mountain Range leeward to the east might well become the source of devastating flood problems the following spring.

In addition to particles on the ground, the presence of small particles in the atmosphere could affect the weather in at least two ways. Particles in the lower atmosphere can act as nuclei for the condensation of water vapor, producing an increase in rain and snow; while particles in the stratosphere can act as a sky shield against the sun, producing colder weather on earth. The immediate effect of particles in the atmosphere was dramatically described by Colonel Lunger in his testimony before the Holifield Committee. On the evening following the 1952 MIKE shot, Colonel Lunger observed "an amber glow along the entire horizon. It was the most artificial thing I have ever seen and sensed in my life. We had displaced many millions of tons of coral debris that had been lifted up to forty and fifty thousand feet by the blast."

The phenomenon described by Colonel Lunger is reminiscent of what has been observed after volcanic eruptions. Among the most famous of these is the eruption of Krakatoa in 1883. The dust thrown up by this eruption caused brilliant sunsets all over the world for the next few years. More important, the dust acted as a shield against the sun's rays, reflecting a relatively small but significant amount of energy back out into space. In France, measurements of the radiation intensities of sunlight were taken which showed a decrease of almost 10 per cent for the next three years. The consequent reduction in the amount of energy absorbed by the earth produced a cooling effect which manifested itself as unusually cold weather in many parts of the world.

The explosion of Krakatoa was the greatest since 1783, when Mt. Asama in Japan produced the most frightful eruption on record. In the Krakatoa eruption, the quantity of earth blown up into the atmosphere amounted to a cubic mile, according to one estimate, most of which descended in the surrounding Straits of Sunda. The fine particles took about three months to reach Western Europe. The world's mean temperature declined by about 1° Fahrenheit in 1884, the year following the Krakatoa eruption. However, probably some of the cold weather can be attributed to causes other than volcanic dust.

A more dramatic example of volcanic activity influencing weather occurred in the early part of the nineteenth century. A series of volcanic eruptions, beginning around 1808 and culminating with the eruption of Mt. Tomboro, on Soembawa Island in Indonesia, on April 7 to 12, 1815, resulted in a memorable cold spell the world over. The Tomboro eruption killed 56,000 people and blew up so much dust that for three days there was darkness up to three hundred miles away. The following year, 1816, became famous in folklore as "eighteen hundred and froze-to-death," "poverty year," and "the year without a summer."

Volcanic dust is probably the most important single factor in determining the fluctuations in world temperature. Other natural factors also contribute significantly, but an envelope of volcanic dust in the upper atmosphere

could theoretically cover in excess of 10^{11} square feet to a depth of one millimeter, i.e., about 3,600 square miles. Although fallout is not deposited uniformly, a cover less than one millimeter is probably still sufficient to exert a pronounced effect. In addition, the assumption that only 10 per cent of the crater material is deposited as local fallout is probably too low.

probably exerts the most profound cooling effect on the earth's surface.[2] W. J. Humphreys, meteorological physicist of the United States Weather Bureau, has stated that since the beginning of reliable meteorological records at the end of the eighteenth century, the temperature of the earth has been lowered by as much as perhaps 1° as a result of volcanic explosions violent enough to put dust into the upper atmosphere. furthermore, if volcanic explosions during this period had been three or four times more numerous than they actually were, the average temperature probably would have been about 2° to 3½° lower than was observed, which is low enough, if continued for a long period of time, to depress the snow line 300 meters (almost 1,000 feet), and thus initiate a moderate ice age.

A reduction of only 10 per cent in the amount of solar radiation striking the earth's surface is sufficient to result in a world-wide cooling of about 11° if maintained over a long period of time. The glaciation that would result from this great a temperature drop would equal or exceed the most extensive experienced in any ice age. The eruption of Mt. Katmai in in 1912 resulted in an over-all 10 per cent reduction of solar radiation. Data obtained on this dust cover permitted Humphreys to calculate that the amount of dust required to cause a 20 per cent reduction in the amount of solar radiation is "astonishingly small"; only about 1/1000 of a cubic mile would suffice to cover the earth.

The amount of material gouged out of the ground by a single twenty-megaton bomb detonated on soil is approximately 1/40 of a cubic mile. This means one hundred such detonations (a total of two thousand megatons) would gouge out somewhat over two cubic miles. If one assumes that 5 per cent of this material were injected into the stratosphere as relatively small particles (less than 20 microns in diameter), then two thousand megatons would inject 1/10 cubic mile into the stratosphere. Of this stratospheric material, perhaps only 1 per cent might consist of very small particles (less than 2 microns in diameter), which are of greatest importance in the cooling process; therefore two thousand megatons could produce about 1/1000 of a cubic mile of fine dust particles, sufficient to reduce the solar radiation by 20 per cent. The reduction in solar radiation caused by the presence of fine atomic ash in the stratosphere would not immediately bring about a change in world-wide temperature, since the oceans act as a buffer. However, over large land masses the effect would probably be more immediate.

The question arises: How long would this cold weather last? This depends not only on the number of particles in the stratosphere, but also on their rate of descent. Physicist R. U. Ayres of the Hudson Institute has calculated that two thirds of the 2 micron particles would have descended by the end of a year and a half. Therefore, a year and a half after a six-thousand-megaton war (three thousand megatons on each side) there would still be sufficient debris in the

[2]It is not certain whether volcanic dust reflects the sun's radiation directly, or whether some more indirect mechanisms might also be involved. Dr. H. Wexler, former Director of Research of the United States Weather Bureau, has suggested that the volcanic dust particles may also serve as nuclei for ice-crystal formation in the atmosphere, causing an increase in the total high-level cloud cover. Other effects are also conceivable. In any case, particles in the air reduce the surface temperature; for example, on September 25 and 26, 1950, the solar radiation received at Washington, D.C., was only 52 per cent of normal, even though it was a cloudless day. This reduction in solar radiation was caused by a pall of smoke emanating from extensive forest fires in western Canada which covered the eastern United States, Canada, and Europe. It was estimated that during the two days the loss of radiant energy lowered the maximum temperature in Washington by 5° to 10° fahrenheit.

stratosphere to reduce the solar radiation by 20 per cent. This implies that the cold spell would last several years, but one can only speculate on the meaning of this cold spell. If the temperature drop averaged out to about 7° over eastern North America, it would mean that the weather in Washington, D.C., would be more like the weather currently observed in Ottawa, Canada.[3] This would have profound implications for agriculture: the wheat belt, for example, would be shifted about five hundred miles south. If, on the other hand, the temperature drop averaged out to about 14°, it is likely that there would be years like 1816, the "year without a summer." A drop of 14° means that in New York City, for example, temperatures in the middle of July would be like those in the middle of May.

Presumably after a few years, as the atomic dust settled, the weather would return to normal. However, there is the remote possibility that the cold spell might last long enough to initiate a vicious cycle. The presence of a white snow cover on the surface of the earth greatly increases the reflection of solar radiation back out into space. Whereas water and land surfaces tend to reflect from 3 to 35 per cent of the solar radiation, a snow cover tends to reflect between 50 and 85 per cent. The snow cover does not appreciably inhibit the release of heat from the surface of the earth out into space. Thus, one can readily envision the mechanism proposed by W. J. Humphreys, whereby the land surfaces heat up barely at all throughout most of the year, while water surfaces near the Arctic tend to freeze up more and more each winter. If a sufficiently large part of the North Atlantic were to freeze over to deflect the Gulf Stream, then Europe would freeze over completely. (England is as far north as Labrador, but the west winds picking up heat from the Gulf Stream keep the continent warm.) Although the advent of an ice age following a thermonuclear war seems unlikely, it is not until the mechanisms of ice-age formation are clearly understood that one can positively rule out this possibility.

If the detonation of a large number of surface bursts should lead to an ice age, then the distortion of nature would last for millennia. A number of other upsets would also follow: the weather pattern in many parts of the world would be altered, the level of the oceans would subside as more and more water became trapped by the glaciers, mass population migrations would alter the current social and political make-up. However, even in the absence of so cataclysmic an event as the creation of an ice age, the cascading of natural upsets following a nuclear disaster might lead to profound changes. The recovery of nature is by no means a foregone conclusion. There is a very real possibility that if the United States were hit hard enough by a nuclear attack most of the country would be converted into a barren desert.

The essence of the problem is this: If the plant cover is destroyed because the radiation-damaged plants succumb to insects, to unusually harsh weather, or to other forces, will the exposed earth erode irreversibly? Under certain circumstances the original plant cover can be restored. Under other circumstances, a plant cover may be restored but may be different from the one that originally characterized the area. A third possibility is that there would be a downward spiral in which various factors would combine to aggravate an already bad situation, producing an environment too hostile to permit the natural re-establishment of a plant cover. Certainly there is sufficient historical experience to indicate that if erosion became a con-

[3]In eastern North America, decreasing the temperature by 1° is equivalent to moving seventy miles north.

tinental, rather than a local, problem the United States would be likely to suffer the same fate as North Africa, the Middle East, and other parts of the world that eroded into desert.

As we have seen, local fallout is likely to trigger a chain of events that would lead to insect plagues, which, of course, would contribute to the destruction of the plant cover. However, insect plagues would not arise if the climate became sufficiently cold. Yet ground bursts, which produce local fallout, are also likely to produce enough stratospheric dust to cause severe cooling. Cold weather slows up the development of many insects in summer and greatly increases insect mortality in winter. Moreover, many radiation-damaged plants, which would serve as a stimulus to an expanding insect population by providing a choice food supply, would probably be killed by the cold weather. For these reasons the insect populations would be drastically reduced by the cold weather and would either not build up at all, or would do so only following the return of normal temperatures. On the other hand, if an insect and its host had a sufficient range, then a large insect population might build up in the south, migrating north as climatic conditions became favorable.

In those parts of the country where cooling effects were serious enough to prevent the build-up of insect populations, radiation-damaged trees and other plants might be unable to cope with the additional stress of unseasonable frosts and shortened growing seasons. The ecologist R. B. Platt has observed in his studies of the irradiated forest in northern Georgia that even under normal conditions in spring radiation-damaged trees break dormancy anywhere from one to eight weeks later than do nonirradiated controls, depending upon the extent of injury. This would mean that many fallout-damaged trees would break dormancy two months later, at a time when the growing season may be perhaps two months shorter.

Radiation, cold weather, and insects offer the most widespread insult to the plant cover, but there would be other serious stresses as well. Some of these, such as disease and fire following insect devastation, have already been discussed in the previous chapter. In addition, there would follow eroision, floods, and drought, while thousands of square miles would probably be destroyed within hours after an attack as a result of the heat and blast released by the nuclear detonations. Blast damage would probably not be very important because blast areas would most likely be too limited to have a significant effect on nature, and most plants have a tremendous capacity to regenerate (a single cell originally derived from the stem of a tobacco plant has been shown to regenerate an entire tobacco plant). This is borne out by studies on atom-blasted islets at Eniwetok. Heat, on the other hand, could ignite very large tracts of forest. These fires would be worse than ordinary forest fires in that a large forest fire takes days and even weeks to cover a large area, whereas a twenty-megaton low air burst would ignite many small fires in a 1,000-square mile area *simultaneously*, resulting in a total burn-out of the area. Even at its peak, the Tillamook fire in Oregon was moving along its fifteen-mile front at an average rate of two miles an hour,[4] a rate of burning that permitted birds and many mammals to get out of the way. Furthermore, large

[4]The Tillamook forest fire burned a total of 486 square miles of virgin Douglas fir. From August 14, 1933, until August 24, the fire burned about 63 square miles. On the 24th, the humidity dropped rapidly to 26 per cent and hot gale-force winds from the east sprang up. During the next twenty hours the fire raged over an additional 420 square miles at a rate of about twenty-one square miles per hour along a fifteen-mile front. The fire stopped as the wind ceased and a thick, wet blanket of fog drifted in from the ocean.

tracts of forest remained untouched: in the very center, a 25,000-acre tract was spared.

How fast is a destroyed plant cover likely to regenerate? In the first place, some initial reproduction of plant life is virtually assured. Certain plants are highly radioresistant and these might initially displace more common, but also more sensitive, plants. Second, the soil constitutes a reservoir of seeds which, because they are underground and because of the inhomogeneity of fallout deposition, would survive. They would need only favorable ecological circumstances to germinate and produce normal plants. These plants would most likely be those currently considered weeds, but they would be admirably suited for establishing a plant cover. This process has actually been observed under natural conditions, and in the course of solving problems associated with erosion. Third, many perennial grasses and ferns, as well as other plants, have shoot meristems which are at or below the ground surface. Additional shielding provided by snow, other plants, or rocks, houses, bridges, and the like would insure the initial survival of some of these meristems capable of vegetative reproduction. Whether the plants coming up from seeds or shielded meristems would prosper and repopulate the earth, or whether they would succumb to assault by insects, microbes, fires, erosion, or adverse climatic events would be determined largely by how rapidly the ecosystem can reconstitute itself.

Normally, if the vegetation is destroyed, nature restores the balance. Terrestrial plant succession after fires, for example, follow some well-established principles. In many parts of the eastern United States, burned-over areas are soon covered by grasses and large annual herbs such as milkweeds, goldenrods, asters and evening primroses. In wetter areas, sedges of various species, water hemlock, blue-eyed grass, and rushes make their appearance. With time, the saplings of shrubs and trees begin to grow. Among them, sumac, aspen, poplar, blackberries, hawthorn, black locust, red and white cedars, and mulberry begin to fill up the open fields. These, in turn, are replaced by taller trees such as red maples and willows in moist areas, or aspens and grey birch, or locusts, oaks, and hickories in dry areas. Ultimately, the oaks and hickories tend to completely dominate the biotic community of plants, the community having once again achieved its stabilized "climax" vegetation. Such a climax community, the culmination of a succession of communities of the type just described, may consist of other trees, such as beech, hemlock, and maple, and it may evolve via different intermediate stages. However, it should be understood that an almost predictable sequence of plant communities arises to restore the former balance in any given location. The return of the climax flora is accompanied by a similar sequence of events tending to restore the climax fauna.

In order for natural repair to occur, other ecological elements must remain essentially intact. A decrease either in rainfall or soil fertility can drastically inhibit the kind of recovery discussed above. A. H. Benton and W. E. Werner, in their book, *Principles of Field Biology and Ecology,* provide an example of a biotic community diverted from its climax forest as a result of a combination of human activities and fire. Between Albany and Schenectady, in eastern New York State, there is an area of sandy soil that covers a strip some twenty miles long and five or six miles wide. Much of this land was once covered with a climax forest of white pine. As a result of agricultural activity, the trees were cut, and parts of the sandy soil were farmed. Fires occurred frequently, burning what little humus had built up, and eventually the whole area became a sandy plain. White pine, which could not tolerate the fires, was replaced by pitch pine, which could. Today the pitch pines grow scattered about as in a savanna, while shrubs of bear oak, dwarf cherries, staghorn, and smooth sumacs comprise the main vegetation underneath.

The most extreme cases of denudation of a plant cover are found around certain ore-smelting towns where the plants are poisoned by toxic products. In one such area, near Copper Hills, Georgia, the release of sulphur dioxide fumes from the copper-smelting operations during the first part of this century denuded the local countryside of its vegetation. The high temperatures and rainfall in the area favored rapid erosion, so that thirty years after the introduction of modern refining methods, which contain the noxious fumes, the area was still barren. Attempts to reforest the area have not yet succeeded. Erosion and the changes in micro-climate that accompany it have combined with the original destructive forces to create a desert in which the land has become too hostile for even artificial reconstruction by conventional techniques.

Thus, the comforting assumption that nature tends to restore its former balance following the destruction of vegetation is not always borne out by experience. In many instances nature arrives instead at a new balance, a balance obviously quite hostile to man.

Nowhere is this more clearly illustrated than in North Africa. Fertile fields once covered most of Egypt, whereas now they are confined almost entirely to the Nile Valley. Beneath the Sahara Desert, covered by thick layers of sand, are traces of dense forest which, less than two thousand years ago, made a rich colony of what is now barren soil and arid land. At El Djem, hidden partly beneath the sands, are the ruins of the great town of Thysdrus, noted for its stadium built to hold 60,000 spectators. Like other buried cities scattered across the North African desert, these ruins are a mute reminder of the fact that this region was once "the granary of the Roman empire." Since Roman times, well over 12,000,000 acres of forest have disappeared in Morocco alone, as a result of fire and overgrazing by sheep and goats.

The cause of the decline of North Africa is popularly attributed to climatic changes, the theory being that the area became hotter and drier and the people were forced to abandon a thriving civilization. However, some geologists, after carefully weighing the evidence, have challenged the conclusion that the climate has changed in any important way since Roman times. Although climatological factors may also have contributed, it was man who destroyed the balance of nature responsible for maintaining soil fertility and moisture.

Why is it that under certain circumstances, the plant cover is unable to regenerate itself, thus setting the stage for serious erosion? To answer this question, let us first take an extreme example, the destruction of tropical rain forests. In general, the equatorial soil is poor: forests can exist in these regions only because they are part of a balanced cycle. All that the forest produces is returned to the forest. The organic matter that falls from the trees constitutes the humus that the forest requires. When man cuts down all the trees to make room for crops, the soil is laid bare and, deprived of shade, heats up. An increase in soil temperature increases the rate of decomposition of organic matter. Organic nitrogen is converted to soluble ammonia and nitrates, which the rains quickly leach away. Roger Bouillene, Director of the Botanical Institute and Garden at the University of Liège, has calculated that a rise in temperature from 77° to 78.8° Fahrenheit may increase the loss of nitrogen by fifteen to twenty pounds per acre per year. With the fertility of the soil destroyed, an irreversible change from forest to desert can be initiated.

Although the tropical rain forests are more susceptible to this type of destruction than are the leafy forests of the temperate zones, the change from forest to grassland to desert has occurred in many parts of the world, largely as a result of man's activities. Vegetation, soil, and microclimate form an interrelated dynamic complex which may be upset by any number of factors directly or indirectly attributable to nuclear war. The destruction of the plant cover, for example, immediately creates a much more hostile environment, an environment which heats up much more when the weather is hot, cools much more when it is cold, and which generally provides much less water. According to a study conducted in the north of Germany, if one simply removed the hedges along the fields, the increased evaporation caused by wind would be great enough to require one third more annual rainfall to compensate for the loss of the hedges. The presence of the hedges probably increased the grain yields by 20 per cent.

In the United States and other parts of the world, extensive use of shelter belts reduces evaporation and also conserves soil moisture by trapping snow and reducing runoff. A canopy of forest trees maintains a higher humidity, lowers high temperatures, and, in general, dramatically reduces the soil water loss by evaporation underneath. For example, a study conducted some years after the Tillamook forest fire compared a tract of forest that had remained intact with the surrounding burn area, which was covered with weeds and shrubs. The study showed that the area covered with weeds and shrubs was 10° hotter on the hottest day, 2° colder on the coldest night, and after ten days of drought showed not only a significantly lower relative humidity in the area but also a lowered soil moisture content. perhaps most important was that the soil of the burn area contained about 20 per cent less organic material, and only about two fifths as much available phosphorus as did the soil of the intact forest.

Other studies have shown that under full shade, soil-surface temperatures remain cooler than air temperatures, even during the hottest part of the day, in contract to unshaded soils, which are always warmer than the air only a few millimeters above ground. The difference between the surface temperature of thinly shaded and exposed dry, gray soil, when measured in the early afternoon of a clear day, exceeded 20°. A black, dry exposed surface was 25° hotter. In winter, on the other hand, dry soil is more prone to frost since the lower its moisture content, the poorer its heat conductivity. Clearly, a dry, exposed soil surface provides a very rigorous microenvironment.

Changing the microclimate in the grasslands can profoundly influence the rate of reproduction of certain insects. Grasshopper plagues originate not where the vegetation is lush, but in areas that form a mosaic of plants and bare soil. Similarly, the European rabbit thrives in areas of depleted, eroded soil. A build-up of these plant-eating animals can lead to overgrazing and the destruction of the remaining plants. Overgrazed plants are less able to cope with draught conditions. This was observed in Nebraska during the dry years in the early 1930's. Native grasses in long-overgrazed prairie, even though protected during recovery, wilted much more than did plants in adjacent prairie that had not been grazed. The leaves rolled or folded, and many of the lower ones dried and lost their green color. Studies showed that overgrazed plants develop poor root systems because whenever 50 per cent or more of the top of a grass plant has been removed, root growth stops completely and will not resume until there has been adequate shoot recovery six to eighteen days later. It is for this reason that overgrazed plants succumb readily to drought.

It is not surprising, therefore, to find that where grasslands were excessively challenged by overgrazing, the grasses were replaced by creosote bushes, mesquite, and cactus, forming a desert community. In Holland, the critical factor in sand-dune erosion proved to be grazing by rabbits. When the rabbits were eliminated as a result of disease (myxomatosis), erosion ceased to be a serious problem and the dune-fixing vegetation suddenly spread dramatically over the dune surfaces. Similarly, in Australia, land that had been previously classed as desert turned into grassland following the elimination of the rabbits.

Fallout, like overgrazing, would either create the kind of mosaic favorable to the build-up of grasshoppers, or it would stress the grasses so that they would be unable to survive severe unseasonable frosts, fire, or drought. This would be particularly true for the marginal grasslands throughout the Mountain and West Central States. Stripping the prairies of their grasses would subject them to wind erosion.[5]

One of America's leading ecologists, Paul B. Sears, has pointed out that "erosion, like many another curse of humanity, grows by what it feeds on." As the wind removes the fertile covering from one field, it destroys the vegetative covering on the next by inundating it with dust. The amount of dust that may be transported by winds is phenomenal. One need recall only the giant dust storms emanating from the dust bowl of the 1930's to appreciate this fact. Early in May 1934, hot, dry weather, followed by a strong wind, created a giant dust cloud about 1,800 miles wide, which struck New York City on May 11. The *New York Times* reported that "yellow dust, driven before a westerly wind, swept over the Eastern Seaboard region from Canada to far south of Washington. This phenomenon, never before experienced in Eastern America, was caused by a wind-driven fog of soil particles that came from drought-stricken areas." One estimate indicated that the cloud contained 300,000,000 tons of topsoil blown off the Central States and sprinkled over half the nation. The fine particles trebled the density of New York's atmosphere, veiling the sun for about five hours and requiring lights to be turned on in the middle of the day.

Serious as the dust-bowl problem was in the 1930's, it might be dwarfed by nuclear-attack dust-bowl problems. Much larger areas could be involved, resulting in an increased frequency and severity of dust storms, and such dust storms might be particularly noxious because the dust would probably contain substantial amounts of long-lived radioactive fallout products, such as radioactive strontium and cesium.

The formation of a dust bowl is in itself by no means an irreversible phenomenon. In some areas of the dust bowl the dust, erosion, and extreme lack of moisture, plus the overgrazing and grasshopper hordes, killed off so much plant life that on ungrazed land it reduced the percentage of plant cover in a five-year period from 90 per cent to about 25 per cent, and on overgrazed land from 80 per cent to almost zero within four years. Yet, between 1940 and 1942 both kinds of land made a complete recovery.

[5]Soils with 80 per cent or more of fine sand are very vulnerable to erosion. Sand soils move readily when wind speeds six inches above ground exceed eleven miles per hour. This occurs when wind velocities at four and a half feet above ground exceed fourteen miles per hour. Winds exceeding these velocities are common in the Great Plains. Furthermore, the relatively high organic-matter content of the Great Plains soil accentuates their susceptibility to wind erosion.

Vigorous efforts can restore the productivity of even dry, infertile land. The state of Israel has conducted a vigorous land-reclamation program, including the draining of about seventy square miles of marshland and the irrigation of about five hundred square miles of arid land. As a result of these efforts, Israel was able to double the land under productive cultivation. The events in this part of the world illustrate the twin problem of deserts and swamps that may be created once unfavorable ecological balances bring about unchecked, accelerated erosion. The Huleh Basin, at the head of the Jordan Valley, was a fertile and thickly populated area in Roman times. This, and other areas at the eastern end of the Mediterranean, began to decline in productivity during the fading of the Byzantine Empire, about 1,300 years ago. The Huleh Basin became a dismal swamp—a focus of malaria infection–as a result of the deposition of sediments from eroding uplands to the north which progressively filled in the northern end of Lake Huleh. The successful reclamation of the area is a triumph of modern technology.

It should be recognized, however, that the Israeli experience may not be applicable after a nuclear attack. The Israelis had not only considerable and skilled manpower, but also very substantial material help from other countries. If much of the economy remained intact after a nuclear attack, the Israeli feat could probably be duplicated in this country. However, if the surviving population were too weak, or too small, and if no outside aid materialized, the country would be unable to check the erosive decay.

Thus far we have considered primarily wind erosion. Equally important is water erosion on hilly or mountainous terrain. Plants on slopes soften the impact of raindrops by intercepting them on leaves or on litter under the plants, while organic matter and roots improve conditions of infiltration and thereby reduce runoff. Many of the factors we discussed earlier in relation to the retention of soil moisture by a plant cover also relate to reducing soil erosion. Plant roots tend to hold soil in place; in their absence rains leach out the minerals and wash the soil off the surface, first downhill, then downstream. Soil not washed off the surface may become compacted by the rain instead. This compaction reduces the ability of the soil to absorb water, resulting in increased runoff which, if sufficiently rapid, causes floods. Furthermore, as we have already seen in connection with the Huleh Basin, deposition is the complement of erosion. The soil particles carried downstream are deposited wherever the water speed diminishes sufficiently for the particles to settle out. This results in the silting up of reservoirs and irrigation systems. In the case of floods, the particles are deposited on the flood banks. At first this deposition actually enriches the valley with the topsoil carried down from upstream. However, if erosion is sufficiently severe, more and more of the soil that is deposited will consist of less fertile subsoil, until finally the flood plains may also become unproductive, having become inundated with infertile subsoil carried down from the destroyed land upstream.

The extent of erosion following a nuclear attack would depend largely on how quickly a devastated area could regenerate a plant cover. Survival of the ecosystem would depend on meteoroligical factors such as wind, precipitation, and temperature, and on geological factors such as the nature of the terrain and of the soil. For example, a flat plain with much of its humus left intact in a not-too-windy area of adequate rainfall would probably reconstitute itself rapidly, returning to its former climax community within one to a few years. On the other hand, if most of the humus were destroyed, either a strong wind or rains on the side of a slope would complete the denuding of the earth's fertile covering. The more unstable ecosystems, such as the marginal grasslands, would therefore be likely to succumb to the stresses placed on them by nuclear war.

An example of erosion culminating in a man-made desert is found in Syria. W. C. Lowdermilk, former Assistant Chief of the Soil Conservation Service, describes the case of "the hundred dead cities." Between Alleppo, Hama, and Antioch there is an area of about 1,500 square miles of rolling limestone country which flourished until the seventh century. Then a series of invasions, first by the Persian Army, then, more disastrously, by nomads, wrought destruction and caused the methods for conserving soil and water to fall into disuse. Today one may still find the ruins of over a hundred former towns, "stark skeletons in beautifully cut stone, standing high on bare rock."

There is a popular misconception that a desert is a place that is very hot all the time. It is true that the barrenness of a desert causes it to heat up more when the sun is shining on it, but a desert can also be a very cold place.[6] The Syrian case is of interest because Alleppo is no farther south than Norfolk, Virginia, and the climate of Syria is comparable to the climate in much of the central and southeastern United States, although the mean annual rainfall is more comparable to Iowa or Illinois.

Meteorological conditions in the United States are really not very different from conditions in those parts of the world that have suffered calamitous erosion. As a matter of fact, some of the worst climate in the world is found over most of the United States. The climate of the West Central and Rocky Mountain States is rugged, and in many areas—from the Badlands in the Dakotas, to the Mojave Desert in southern California—the climate is frequently clearly hostile. East of the Rocky Mountains changes in temperature of up to 50° within a twelve-hour period are not uncommon, as warm tropical air masses coming from the south are displaced by cold polar air masses sweeping in from Arctic Canada. Much of this variability stems from the fact that, unlike Europe and North Africa, North America has no east-west mountain ranges to deflect these air masses. E. R. Biel, former Chairman of the Department of Meteorology at Rutgers University, has pointed out that even in the Northeast "the tropical heat conditions of our summers" cause the water requirements of plants to be very high. Biel points out, however, that we are fortunate that the tropical air masses from the south generally consist of marine air carrying enormous masses of precipitable water. This means that much of the United States is spared those desiccating and dust-laden air masses which are such a liability on other continents.

However, a nuclear attack could cancel out this fortuitous situation, as large areas of the United States became stripped of their plant cover. There is evidence to indicate that clouds that form over deserts are less likely to shed their moisture than are clouds formed over grasslands.[7] Thus, a vicious cycle could be initiated: land stripped of its plant cover as a result of nuclear attack would not only lose its fertility, but might also be deprived of adequate moisture and therefore have little chance to recover.

Mesopotamia stands out as the classical

[6]For example, the Great Basin Desert that extends north from southern Nevada and southern Utah may be subjected to killing frosts any month of the year; and in fall, winter, and spring frosts occur almost every night, often accompanied by bitter cold.

[7]Although this evidence is far from conclusive, it is supported by other observations, also not conclusive as yet, which indicate that the presence of forests actually increase annual rainfall.

model of irreversible destruction visited upon a society by ecological devastation following a fatal war. For over 6,000 years the valleys of the Tigris and the Euphrates were a spawning ground of human civilization and progress. This was true in spite of the fact that Mesopotamia was repeatedly invaded and conquered. Babylonians, Assyrians, Persians, Macedonians, Parthians, Romans, and others, conquered part or all of the region. Then, in the middle of the thirteenth century came the Mongol invasion under Hulagu Khan, grandson of Genghis Khan, which, unlike earlier conquests, succeeded in destroying the country. The week after the surrender of Baghdad, 800,000 people were put to the sword. The earthen works on which the irrigation system depended were destroyed, and the canal system on which the land depended fell into ruin. As usual, war was followed by famine and epidemic diseases, which, coupled with the initial slaughter, left the survivors too weak to repair the complex irrigation system. The area that had been so productive rapidly turned into desert, and there followed a continuing struggle between the city and the nomads who roamed the impoverished countryside. Although millions of Mesopotamians survived, the "cradle of civilization" was no longer significant to the mainstream of human advancement. Mesopotamia's fate is in some ways most similar to what may be expected as a result of nuclear attack. Because of the staggering loss of both lives and resources, ecological destruction could not be checked, and a technologically advanced society was permanently destroyed.

6.

Physics in the Contemporary World

J. Robert Oppenheimer

J. Robert Oppenheimer has been director of the Institute for Advanced Study at Princeton, New Jersey. He is a world-renowned physicist trained at Harvard, Cambridge, and Göttingen. He was Professor of Physics at the University of California and the California Institute of Technology for eighteen years. Between 1943 and 1945 he was Director of the laboratory at Los Alamos, where the first atomic bombs were made. From 1945 through 1953 he served in the many advisory positions to the Atomic Energy Commission, the White House, and the Departments of State and Defense. In the spring of 1954, in a much-publicized proceeding, he was denied security clearance.

His books include *The Open Mind* and *Science and the Common Understanding.*

If I have even in the title of this talk sought to restrict its theme, that does not imply an overestimate of physics among the sciences, nor a too great myopia for these contemporary days. It is rather that I must take my starting point in the science in which I have lived and worked, and a time through which my colleagues and I are living.

Nevertheless, I shall be talking tonight about things which are quite general for the relations between science and civilization. For it would seem that in the ways of science, its practice, the peculiarities of its discipline an universality, there are patterns which in the past have somewhat altered, and in the future may greatly alter, all that we think about the world and how we manage to live in it. What I shall be able to say of this will not be rich in exhortation, for this is ground that I know how to tread only very lightly.

But that I should be speaking of such general and such difficult questions at all reflects in the first instance a good deal of self-consciousness on the part of physicists. This self-consciousness is in part a result of the highly critical traditions which have grown up in physics in the last half century, which have shown in so poignant a way how much the applications of science determine our welfare and that of our fellows, and which have cast in doubt that traditional optimism, that confidence in progress, which have characterized Western culture since the Renaissance.

It is, then, *about* physics rather than *of* physics that I shall be speaking—and there is a great deal of difference. You know that when a student of physics makes his first acquaintance with the theory of atomic structure and of quanta, he must come to understand the rather deep and subtle notion which has turned out to be the clue to unraveling that whole domain of physical experience. This is the notion of complementarity, which recognizes that various ways of talking about physical experience may each have validity, and may each be necessary for the adequate description of the physical world, and may yet stand in a mutually exclusive rela-

 This lecture was delivered on November 25, 1947 as the second Arthur Dehon Little Memorial Lecture at the Massachusetts Institute of Technology in Cambridge.

tionship to each other, so that to a situation to which one applies, there may be no consistent possibility of applying the other. Teachers very often try to find illustrations, familiar from experience, for relationships of this kind; and one of the most apt is the exclusive relationship between the practicing of an art and the description of that practice. Both are a part of civilized life. But an analysis of what we do and the doing of it—these are hard to bed in the same bed.

As it did on everything else, the last world war had a great and at least a temporarily disastrous effect on the prosecution of pure science. The demands of military technology in this country and in Britain, the equally overriding demands of the Resistance in much of Europe, distracted the physicists from their normal occupations, as they distracted most other men.

We in this country, who take our wars rather spastically, perhaps witnessed a more total cessation of true professional activity in the field of physics, even in its training, than did any other people. For in all the doings of war we, as a country, have been a little like the young physicist who went to Washington to work for the National Defense Research Committee in 1940. There he met his first Civil Service questionnaire and came to the questions on drinking: "Never," "occasionally," "habitually," "to excess." He checked both "occasionally" and "to excess." So, in the past, we have taken war.

All over the world, whether because of the closing of universities, or the distractions of scientists called in one way or another to serve their countries, or because of devastation and terror and attrition, there was a great gap in physical science. It has been an exciting and an inspiring sight to watch the recovery—a recovery testifying to extraordinary vitality and vigor in this human activity. Today, barely two years after the end of hostilities, physics is booming.

One may have gained the impression that this boom derives primarily from the application of the new techniques developed during the war, such as the atomic reactor and microwave equipment; one may have gained the impression that in large part the flourishing of physics lies in exploitation of the eagerness of governments to promote it. These are indeed important factors. But they are only a small part of the story. Without in any way deprecating the great value of wartime technology, one nevertheless sees how much of what is today new knowledge can trace its origin directly, by an orderly yet imaginative extension, to the kind of things that physicists were doing in their laboratories and with their pencils almost a decade ago.

Let me try to give a little more substance to the physics that is booming. We are continuing the attempt to discover, to identify and characterize, and surely ultimately to order, our knowledge of what the elementary particles of physics really are. I need hardly say that in the course of this we are learning again how far our notion of *elementarity*, of what makes a particle elementary, is from the early atomic ideas of the Hindu and Greek atomists, or even from the chemical atomists of a century ago. We are finding out that what we are forced to call elementary particles retain neither permanence nor identity, and they are elementary only in the sense that their properties cannot be understood by breaking them down into subcomponents. Almost every month has surprises for us in the findings about these particles. We are meeting new ones for which we are not prepared. We are learning how poorly we had identified the properties even of our old friends among them. We are seeing what a challenging job the ordering of this experience is likely to be, and what a strange world we must enter to find that order.

In penetrating into this world perhaps our sharpest tool in the past has been the observation of the phenomena of the cosmic rays in interaction with matter. But the next years will see an important methodological improvement, when the great program of ultra-high-energy accelerators begins to get under way. This program is itself one of the expensive parts of physics. It has been greatly subsidized by the government, primarily through the Atomic Energy Commission and the Office of Naval Research. It is a superlative example, of which one could find so many, of the repayment that technology makes to basic science, in providing means whereby our physical experience can be extended and enriched.

Another progress is the refinement of our knowledge of the behavior of electrons within atomic systems, a refinement which on the one hand is based on the microwave techniques, to the developments of which the Radiation Laboratory of the Massachusetts Institute of Technology made unique contributions, and which on the other hand has provided a newly vigorous criterion for the adequacy of our knowledge of the interactions of radiation and matter. Thus we are beginning to see in this field at least a partial resolution, and I am myself inclined to think rather more than that, of the paradoxes that have plagued the professional physical theorists for two decades.

A third advance in atomic physics is in the increasing understanding of those forces which give to atomic nuclei their great stability, and to their transmutations their great violence. It is the prevailing view that a true understanding of these forces may well not be separable from the ordering of our experience with regard to elementary particles, and that it may also turn on an extension to new fields of recent advances in electrodynamics.

However this may be, all of us who are physicists by profession know that we are embarked on another great adventure of exploration and understanding, and count ourselves happy for that.

In how far is this an account of physics in the United States only? In how far does it apply to other parts of the world, more seriously ravaged and more deeply disturbed by the last war? That question may have a somewhat complex answer, to the varied elements of which one may pay respectful attention.

In much of Europe and Japan, that part of physics which does not rest on the availability of elaborate and radical new equipment is enjoying a recovery comparable to our own. The traditional close associations of workers in various countries makes it just as difficult now to disentangle the contributions by nationality as it was in the past. But there can be little doubt that it is very much harder for a physicist in France, for instance, or the Low Countries, and very much more nearly impossible for him in Japan, to build a giant accelerator than it is for the workers in this country.

Yet in those areas of the world where science has not merely been disturbed or arrested by war and by terror, but where terror and its official philosophy have, in a deep sense, corrupted its very foundations, even the traditional fraternity of scientists has not proved adequate protection against decay. It may not be clear to us in what way and to what extent the spirit of scientific inquiry may come to apply to matters not yet and perhaps never to be part of the domain of science; but that is does apply, there is one very brutal indication. Tyranny, when it gets to be absolute, or when it tends so to become, finds it impossible to continue to live with science.

Even in the good ways of contemporary physics, we are reluctantly made aware of our dependence on things which lie outside our science. The experience of the war, for those who were called upon to serve the survival of their civilization through the Resistance, and for those who contributed more remotely, if far more decisively, by the development of new instruments and weapons of war, has left us with a legacy of concern. In these troubled times it is not likely that we shall be free of it altogether. Nor perhaps is it right that we should be.

Nowhere is this troubled sense of responsibility more acute, and surely nowhere has it been more prolix, than among those who participated in the development of atomic energy for military purposes. I should think that most historians would agree that other technical developments, notably radar, played a more decisive part in determining the outcome of this last war. But I doubt whether that participation would have of itself created the deep trouble and moral concern which so many of us who were physicists have felt, have voiced, and have tried to get over feeling. It is not hard to understand why this should be so. The physics which played the decisive part in the development of the atomic bomb came straight out of war laboratories and our journals.

Despite the vision and the far-seeing wisdom of our wartime heads of state, the physicists felt a peculiarly intimate responsibility for suggesting, for supporting, and in the end, in large measure, for achieving the realization of atomic weapons. Nor can we forget that these weapons, as they were in fact used, dramatized so mercilessly the inhumanity and evil of modern war. In some sort of crude sense which no vulgarity, no humor, no overstatement can quite extinguish, the physicists have known sin; and this is a knowledge which they cannot lose.

Probably in giving expression to such feelings of concern most of us have belabored the influence of science on society through the medium of technology. This is natural, since the developments of the war years were almost exclusively technological, and since the participation of academic scientists forced to be deeply aware of an activity of whose existence they had always known but which had been often remote from them.

When I was a student at Göttingen twenty years ago, there was a story current about the great mathematician Hilbert, who perhaps would have liked, had the world let him, to have thought of his science as something independent of worldly vicissitudes. Hilbert had a colleague, an equally eminent mathematician, Felix Klein, who was certainly aware, if not of the dependence of science generally on society, at least of the dependence of mathematics on the physical sciences which nourish it and give it application. Klein used to take some of his students to meet once a year with the engineers of the Technical High School in Hanover. One year he was ill and asked Hilbert to go in his stead, and urged him, in the little talk that he would give, to try to refute the then prevalent notion that there was a basic hostility between science and technology. Hilbert promised to do so; but when the time came a magnificent absent-mindedness led him instead to speak his own mind: "One hears a good deal nowadays of the hostility between science and technology. I don't think that is true, gentlemen. I am quite sure that it isn't true, gentlemen. It almost certainly isn't true. It really can't be true. *Sie haben ja gar nichts mit einander zu tun.* They have nothing whatever to do with one another." Today the wars and the troubled times deny us the luxury of such absent-mindedness.

The great testimony of history shows how often in fact the development of science has emerged in response to technological and even economic needs, and how in the economy of social effort, science, even of the most abstract and recondite kind, pays for itself again and again in providing the basis for radically new technological developments. In fact, most people—when they think of science as a good thing, when they think of it as worthy of encouragement, when they are willing to see their governments spend substance upon it, when they greatly do honor to men who in science have attained some eminence —have in mind that the conditions of their life have been altered just by such technology, of which they may be reluctant to be deprived.

The debt of science to technology is just as great. Even the most abstract researches owe their very existence to things that have taken place quite outside of science, and with the primary purpose of altering and improving the conditions of man's life. As long as there is a healthy physics, this mutual fructification will surely continue. Out of its work there will come in the future, as so often in the past, and with an apparently chaotic unpredictability, things which will improve man's health, ease his labor, and divert and edify him. There will come things which, properly handled, will shorten his working day and take away the most burdensome part of his effort, which will enable him to communicate, to travel, and to have a wider choice both in the general question of how he is to spend his life and in the specific question of how he is to spend an hour of his leisure. There is no need to belabor this point, nor its obverse—that out of science there will come, as there has in this last war, a host of instruments of destruction which will facilitate that labor, even as they have facilitated all others.

But no scientist, no matter how aware he may be of these fruits of his science, cultivates his work, or refrains from it, because of arguments such as these. No scientist can hope to evaluate what his studies, his researches, his experiments may in the end produce for his fellow men, except in one respect—if they are sound, they will produce knowledge. And this deep complementarity between what may be conceived to be the social justification of science and what is for the individual his compelling motive in its pursuit makes us look for other answers to the question of the relation of science to society.

One of these is that the scientist should assume responsibility for the fruits of his work. I would not argue against this, but it must be clear to all of us how very modest such assumption of responsibility can be, how very ineffective it has been in the past, how necessarily ineffective it will surely be in the future. In fact, it appears little more than exhortation to the man of learning to be properly uncomfortable, and, in the worst instances, is used as a sort of screen to justify the most casual, unscholarly and, in the last analysis, corrupt intrusion of scientists into other realms of which they have neither experience nor knowledge, nor the patience to obtain them.

The true responsibility of a scientist, as we all know, is to the integrity and vigor of his science. And because most scientists, like all men of learning, tend in part also to be teachers, they have a responsibility for the communication of the truths they have found. This is at least a collective if not an individual responsibility. That we should see in this any insurance that the fruits of science will be used for man's benefit, or denied to man when they make for his distress or destruction, would be a tragic naïveté.

There is another side of the coin. This is the question of whether there are elements in the way of life of the scientist which need not be restricted to the professional, and which have hope in them for bringing dignity and courage and serenity to other men. Science is not all of the life of reason; it is a part of it. As such, what can it mean to man?

Perhaps it would be well to emphasize that I am talking neither of wisdom nor of an elite of scientists, but precisely of the kind of work and thought, of action and discipline, that makes up the everyday professional life of the scientist. It is not of any general insight into human affairs that I am talking. It is not the kind of thing we recognize in our greatest statesmen, after long service devoted to practical affairs and to the public interest. It is something very much more homely and robust than that. It has in it the kind of beauty that is inseparable from craftsmanship and form, but it has in it also the vigor that we rightly associate with the simple, ordered lives of artisans or of farmers, that we rightly associate with lives to which limitations of scope, and traditional ways, have given robustness and structure.

Even less would it be right to interpret the question of what there is in the ways of science that may be of general value to mankind in terms of the creation of an elite. The study of physics, and I think my colleagues in the other sciences will let me speak for them too, does not make philosopher-kings. It has not, until now, made kings. It almost never makes fit philosophers—so rarely that they must be counted as exceptions. If the professional pursuit of science makes good scientists, if it makes men with a certain serenity in their lives, who yield perhaps a little more slowly than others to the natural corruptions of their time, it is doing a great deal, and all that we may rightly ask of it. For if Plato believed that in the study of geometry, a man might prepare himself for wisdom and responsibility in the world of men, it was precisely because he thought so hopefully that the understanding of men could be patterned after the understanding of geometry. If we believe that today, it is in a much more recondite sense, and a much more cautious one.

Where, then, is the point? For one thing, it is to describe some of the features of the professional life of the scientist, which make of it one of the great phenomena of the contemporary world. Here again I would like to speak of physics; but I have enough friends in the other sciences to know how close their experience is to ours. And I know too that despite profound differences in method and technique, differences which surely are an appropriate reflection of the difference in the areas of the world under study, what I would say of physics will seem familiar to workers in other disparate fields, such as mathematics or biology.

What are some of these points? There is, in the first instance, a total lack of authoritarianism, which is hard to comprehend or to admit unless one has lived with it. This is accomplished by one of the most exacting of intellectual disciplines. In physics the worker learns the possibility of error very early. He learns that there are ways to correct his mistakes; he learns the futility of trying to conceal them. For it is not a field in which error awaits death and subsequent generations for verdict—the next issue of the journals will take care of it. The refinement of techniques for the prompt discovery of error serves as well as any other as a hallmark of what we mean by science.

In any case, it is an area of collective effort in which there is a clear and well-defined community whose canons of taste and order simplify the life of the practitioner. It is a field in which the technique of experiment has given an almost perfect harmony to the balance between thought and action. In it we learn, so frequently that we could almost become accustomed to it, how vast is the novelty of the world, and how much even the physical world transcends in delicacy and in balance the limits of man's prior imaginings. We learn that views may be useful and inspiriting although they are not complete. We come to have a great caution in all assertions of totality, of finality or absoluteness.

In this field quite ordinary men, using what are in the last analysis only the tools which are generally available in our society, manage to unfold for themselves and all others who wish to learn, the rich story of one aspect of the physical world, and of man's experience. We learn to throw away those instruments of action and those modes of description which are not appropriate to the reality we are trying to discern, and in this most painful discipline, find ourselves modest before the world.

The question which is so much in our mind is whether a comparable experience, a comparable discipline, a comparable community of interest, can in any way be available to mankind at large. I suppose that all the professional scientists together number some one one-hundredth of a per cent of the men of the world—even this will define rather generously what we mean by scientists. Scientists as professionals are, I suppose, rather sure to constitute a small part of our people.

Clearly, if we raise at all this question that I have raised, it must be in the hope that there are other areas of human experience that may be discovered or invented or cultivated, and to which the qualities which distinguish scientific life may be congenial and appropriate. It is natural that serious scientists, knowing of their own experience something of the quality of their profession, should just today be concerned about its possible extension. For it is a time when the destruction and the evil of the last quarter century make men everywhere eager to seek all that can contribute to their intellectual life, some of the order and freedom and purpose which we conceive the great days of the past to have. Of all intellectual activity, science alone has flourished in the last centuries, science alone has turned out to have the kind of universality among men which the times require. I shall be disputed in this; but it is near to truth.

If one looks at past history, one may derive some encouragement for the hope that science, as one of the forms of reason, will nourish all of its forms. One may note how integral the love and cultivation of science were with the whole awakening of the human spirit which characterized the Renaissance. Or one may look at the late seventeenth and eighteenth centuries in France and England and see what pleasure and what stimulation the men of that time derived from the growth of physics, astronomy and mathematics.

What perhaps characterizes these periods of the past, which we must be careful not to make more heroic because of their remoteness, was that there were many men who were able to combine in their own lives the activities of a scientist with activities of art and learning and politics, and were able to carry over from the one into the others this combination of courage and modesty which is the lesson that science always tries to teach to anyone who practices it.

And here we come to a point we touched earlier. It is very different to hear the results of science, as they may be descriptively or even analytically taught in a class or in a book or in the popular talk of the time; it is very different to hear these and to participate even in a modest way in the actual attainment of new knowledge. For it is just characteristic of all work in scientific fields that there is no authority to whom to refer, no one to give canon, no one to blame if the picture does not make sense.

Clearly these circumstances pose a question of great difficulty in the field of education. For if there is any truth in the views that I have outlined, there is all the difference in the world between hearing about science or its results and sharing in the experience of the scientist himself and of that of the scientific community. We all know that an awareness of this, and an awareness of the value of science as method, rather than science as doctrine, underlies the practices of teaching to scientist and layman alike. For surely the whole notion of incorporating a laboratory in a high school or college is a deference to the belief that not only what the scientist finds but how he finds it is worth learning and teaching and worth living through.

Yet there is something fake about all this. No one who has had to do with elementary instruction can have escaped a sense of artificiality in the way in which students are led, by the calculations of their instructors, to follow paths which will tell them something about the physical world. Precisely that groping for what is the appropriate experiment, what are the appropriate terms in which to view subtle or complex phenomena, which are the substance of scientific effort, almost inevitably are distilled out of it by the natural patterns of pedagogy. The teaching of science to laymen is not wholly a loss; and here perhaps physics is an atypically bad example. But surely they are rare men who, entering upon a life in which science plays no direct part, remember from their early courses in physics what science is like or what it is good for. The teaching of science is at its best when it is most like an apprenticeship.

President Conant, in his sensitive and thoughtful book *On Understanding Science,* has spoken at length of these matters. He is aware of how false it is to separate scientific theory from the groping, fumbling, tentative efforts which lead to it. He is aware that it is science as method and not as doctrine which we should try to teach. His basic suggestion is that we attempt to find, in the history of our sciences, stories which can be recreated in the instruction and experiment of the student and which thus can enable him to see at firsthand how error may give way to less error, confusion to less confusion, and bewilderment to insight.

The problem that President Conant has here presented is indeed a deep one. Yet he would be quite willing, I think, that I express skepticism that one can re-create the experience of science as an artifact. And he would no doubt share my concern that science so taught would be corrupt with antiquarianism. It was not antiquarianism but a driving curiosity that inspired in the men of the Renaissance their deep interest in classical culture.

For it is in fact difficult, almost to the point of impossibility, to re-create the climate of opinion in which substantial errors about the physical world, now no longer entertained, were not only held but were held unquestioned as part of the obvious mode of thinking about reality. It is most difficult to do because in all human thought only the tiniest fraction of our experience is in focus, and because to this focus a whole vast unanalyzed account of experience must be brought to bear. Thus I am inclined to think that, with exceptions I hope will be many but fear will be few, the attempt to give the history of science as a living history will be far more difficult than either to tell of the knowledge that we hold today or to write externally of that history as it may appear in the learned books. It could easily lead to a sort of exercise of mental inventiveness on the part of teachers and students alike which is the very opposite of the candor, the "no holds barred" rules of Professor Bridgman, that characterize scientific understanding at its best.

If I am troubled by President Conant's suggestions, this is not at all because I doubt that the suggestions he makes are desirable. I do have a deep doubt as to the extent to which they may be practical. There is something irreversible about acquiring knowledge; and the simulation of the search for it differs in a most profound way from the reality. In fact, it would seem that only those who had some firsthand experience in the acquisition of new knowledge in some disciplined field would be able truly to appreciate how great the science of the past has been, and would be able to measure those giant accomplishments against their own efforts to penetrate a few millimeters farther into darkness that surrounds them.

Thus it would seem at least doubtful that the spiritual fruits of science could be made generally available, either by the communication of its results, or by the study of its history, or by the necessarily somewhat artificial re-enactment of its procedures. Rather it would seem that there are general features of the scientists' work the direct experience of which in any context could contribute more to this end. All of us, I suppose, would list such features and find it hard to define the words which we found it necessary to use in our lists. But on a few, a common experience may enable us to talk in concert.

In the first instance the work of science is co-operative; a scientist takes his colleagues as judges, competitors and collaborators. That does not mean, of course, that he loves his colleagues; but it gives him a way of living with them which would be not without its use in the contemporary world. The work of science is discipline in that its essential inventiveness is most of all dedicated to means for promptly revealing error. One may think of the rigors of mathematics and the virtuosity of physical experiment as two examples. Science is disciplined in its rejection of questions that cannot be answered and in its grinding pursuit of methods for answering all that can. Science is always limited, and is in a profound sense unmetaphysical, in that it necessarily bases itself upon the broad ground of common human experience, tries to refine it within narrow areas where progress seems possible and exploration fruitful. Science is novelty and change. When it closes it dies. These qualities constitute a way of life which of course does not make wise men from foolish, or good men from wicked, but which has its beauty and which seems singularly suited to man's estate on earth.

If there is to be any advocacy at all in this talk, it would be this: that we be very sensitive to all new possibilities of extending the techniques and the patterns of science into other areas of human experience. Even in saying this we must be aware how slow the past development of science has in fact been, how much error there has been, and how much in it that turned out to be contrary to intellectual health or honesty.

We become fully aware of the need for caution if we look for a moment at what are called the social problems of the day and try to think what one could mean by approaching them in the scientific spirit, of trying to give substance, for example, to the feeling that a society that could develop atomic energy could also develop the means of controlling it. Surely the establishment of a secure peace is very much in all our minds. It is right that we try to bring reason to bear on an understanding of this problem; but for that there are available to us no equivalents of the experimental techniques of science. Errors of conception can remain undetected and even undefined. No means of appropriately narrowing the focus of thinking is known to us. Nor have we found good avenues for extending or deepening our experience that bears upon this problem. In short, almost all the preconditions of scientific activity are missing, and in this case, at least, one may have a melancholy certainty that man's inventiveness will not rapidly provide them. All that we have from science in facing such great questions is a memory of our professional life, which makes us somewhat skeptical of other people's assertions, somewhat critical of enthusiasms so difficult to define and to control.

Yet the past century has seen many valid and inspiring examples for the extension of science to new domains. As even in the case of physics, the initial steps are always controversial; probably we should not as a group be unanimous in saying which of these extensions were hopeful, and which not, for the science of the future. But one feature which I cannot fail to regard as sound—particularly in the fields of biology and psychology— is that they provide an appropriate means of correlating understanding and action, and involve new experimental procedures in terms of which a new conceptual apparatus can be defined; above all, they give us means of detecting error. In fact, one of the features which must arouse our suspicion of the dogmas some of Freud's followers have built up on the initial brilliant works of Freud is the tendency toward a self-sealing system, a system, that is, which has a way of almost automatically discounting evidence which might bear adversely on the doctrine. The whole point of science is to do just the opposite: to invite the detection of error and to welcome it. Some of you may think that in another field a comparable system has been developed by the recent followers of Marx.

Thus we may hope for an ever-widening and more diverse field of application of science. But we must be aware how slowly these things develop and how little their development is responsive to even the most desperate of man's needs. For me it is an open question, and yet not a trivial one, whether in a time necessarily limited by the threats of war and of chaos these expanding areas in which the scientific spirit can flourish may yet contribute in a decisive way to man's rational life.

I have had to leave this essential question unanswered: I am not at all proud of that. In lieu of apology perhaps I may tell a story of another lecturer, speaking at Harvard, a few miles from here, two decades ago. Bertrand Russell had given a talk on the then new quantum mechanics, of whose wonders he was most appreciative. He spoke hard and earnestly in the New Lecture Hall. And when he was done, Professor Whitehead, who presided, thanked him for his efforts, and not least for "leaving the vast darkness of the subject unobscured."

7.

Danger to the Individual

Edward Teller and Albert L. Latter

Professor Edward Teller is perhaps best known for his original work on the hydrogen bomb. During and after World War II he worked on the Manhattan Project and at Los Alamos. Currently he is a member of the faculty of the University of California at Berkeley.

Dr. Albert Latter, his coauthor, is presently with the Rand Corporation in California doing research in the field of theoretical physics.

Their book, *Our Nuclear Future,* frankly discusses the facts, dangers, and opportunities of nuclear energy. They explain clearly and simply the effects of radioactivity and fallout from the bombs.

How much harm is being done by the atomic tests? Some scientists have claimed that from past tests alone about 50,000 persons throughout the world will die prematurely. There is no general agreement on this point. Some think the number should be smaller. It is possible that radioactivity produces some effects which prolong life rather than shorten it. But even if all the biological consequences of radiation were known many questions would still demand answers. Can tests be justified if they actually shorten some human lives? Even the possibility of a health hazard must be taken most seriously. On the other hand: Are there any reasons which make continued testing necessary?

We shall return to these questions in a later chapter. First, however, we shall try to put before the reader the known facts about the fallout danger to the individual. We shall try to put this danger into perspective by relating it to other more familiar dangers to which all of us are exposed. In the following chapter we shall discuss how the fallout may affect future generations.

The dangers from big doses of radiation are well known. Exposure to a thousand roentgens over our whole body causes almost certain death in less than thirty days. Four or five hundred roentgens give a fifty-fifty chance of survival. At less than a hundred roentgens, there is no danger of immediate death. Three years ago the Marshallese got a dose of 175 roentgens. None died. Apparently all are in good health.

Over longer periods of time even bigger radiation doses can be tolerated. A thousand roentgens spread over a lifetime produce no apparent biological consequences in individual cases. A rough rule (which is not too well-established) is that five times as much radiation can be tolerated if one is exposed to only a little radiation at any one time.

A hundred roentgens all at once, or several times this amount over a protracted time period, will not cause sickness or death that can be directly blamed on the radiation. However, such a dose of radiation may have harmful biological consequences which are more subtle. An exposed individual may develop an increased susceptibility to certain diseases, notably bone cancer and leukemia. Leukemia is a fatal disease in which the white blood cells multiply too rapidly.

A person who receives a hundred roentgens does not necessarily contract bone cancer or leukemia. Rather, his chance of contracting these diseases during his lifetime may have been increased. Knowledge of this kind can be obtained only with the help of statistics.

If, for example, a large number of mice receive a heavy dosage of radiation over a long period of time, one finds that the incidence of tumors and leukemia is higher amongst such irradiated animals than the natural incidence of these diseases.

Direct evidence with human beings—fortunately—is rather scarce. Statistics exist on the survivors of Hiroshima and Nagasaki, and also on radiologists. The latter group probably receive several hundred roentgens during their professional lifetimes. In addition, some statistics exist on children who have been treated with large doses of radiation for enlarged thymuses. Persons suffering from ankylosing spondylitis, which is a painful disease of the spinal joints, have also been treated with large X-ray doses. The statistics in all these cases lead to the same conclusion: that large doses of radiation increase the likelihood that an individual's life will be shortened by leukemia and possibly also other cancers. Furthermore, it appears (mainly from the experiments on animals) that the increased likelihood is simply proportional to the amount of radiation received, at least for doses in the neighborhood of several hundred roentgens or so.

This of course sounds frightening. But the radiation doses from the world-wide fallout are in a completely different class from those we have been discussing. They are very much smaller. On the average human bones are getting about 0.002 roentgens per year from the Sr^{90} in the fallout. In addition the whole body is receiving a roughly equal amount in gamma rays, mainly from Cs^{137}. These figures apply to new bone in young children who have grown up in an environment of Sr^{90} in the northern part of the United States. This is a region of maximum fallout. Adults whose bones were made for the most part before the atomic testing started are getting about 0.0003 roentgens per year from Sr^{90}. None of these figures appears to be alarming.

At this present rate a lifetime dosage in northern U.S. is only a small fraction of a roentgen. A rare individual might get several times this amount. If tests continue at the present rate, radiation levels could increase by as much as fivefold. However, even in this situation it is difficult to imagine anyone receiving a lifetime dose of more than five or ten roentgens from the world-wide fallout. A more reasonable estimate for the average lifetime dose would be a few roentgens or less.

One might conclude from these figures that there is no danger whatsoever from the fallout. This conclusion, however, may not be correct.

The danger from such small doses of radiation is not easy to define. Even the best statistical methods are insufficient. One is looking for small effects which show up only after millions of cases have been studied. Animal experiments are extremely difficult to carry out under these conditions. Direct controlled experience with human beings is, of course, impossible. As a result, one is forced to draw conclusions from the effects at higher dose levels, where experimental data have been obtained.

This may be done in many ways. One way is to assume that the law of proportionality holds down to the smallest doses. This means that one roentgen produces one hundredth as many cases of bone cancer and leukemia as 100 roentgens produce. This law is plausible. It is by no means proven.

By arguing in this way one finds that for each megaton of fission energy which escapes from the test site in the world-wide fallout the lives of approximately four hundred persons would be shortened by leukemia or bone cancer. Under present conditions of testing, roughly one half of the fission products are deposited as close-in fallout in and near the test site. Per megaton of fission energy exploded, therefore, perhaps 200 persons may get leukemia or bone cancer. This figure could actually be higher, possibly even a thousand persons or more per megaton. It could also be lower. It could be zero.

It is possible that radiation of less than a certain intensity does not cause bone cancer or leukemia at all. In the past small doses of radiation have often been regarded as beneficial. This was not supported by any scientific evidence. Today many well-informed people believe that radiation is harmful even in the smallest amounts. This statement has been repeated in an authoritative manner. Actually there can be little doubt that radiation hurts the individual cell. But a living being is a most complex thing. Damage to a small fraction of the cells might be beneficial to the whole organism. Some experiments on mice seem to show that exposure to a little radiation increases the life expectancy of the animals. Scientific truth is firm—when it is complete. The evidence of what a little radiation will do to a complex animal like a human being is in an early and uncertain state.

In any event the number of additional cases of leukemia and bone cancer due to the fallout radiation is certainly too small to be noticed against the natural incidence of these disorders.

In the next thirty years about 6,000,000 people throughout the world will die from leukemia and bone cancer. From past tests, which have involved the explosion of about fifty megatons of fission energy, the possibility exists that another 50 × 200, i.e., 10,000 cases may occur. Statistical methods are not able to find the difference between 6,000,000 and 6,010,000. There is no way to differentiate between the fallout-induced cases of leukemia and bone cancer, and those which occur naturally.

The possible shortening of ten thousand lives may seem rather ominous. But mere figures can be misleading. A better way to appreciate the danger from fallout is to compare it with other more familiar dangers. Such a comparison can be made with the natural background of cosmic rays and radioactivity in the earth and in our own bodies.

We are constantly and inescapably exposed to this radiation. Our ancestors have been exposed to it. The human race has evolved in such a radioactive environment. Moreover, the biological effects from different kinds of radiation can be compared in a meaningful way in terms of roentgens. Therefore the danger from Sr^{90} is not unknown in every respect. In some ways it is very well-known because we and all living beings have spent our days in a similarly dangerous surrounding. We live on an earth which has radioactivity in its rocks, which carries a similar activity in its waters, and which is exposed from all sides, to a rain of particles which produce effects identical with the effects of radioactive materials.

Not all radiations which have the same intensity (the same number of roentgens) have precisely the same effect. The damage produced also depends somewhat on the spacing of the ionized and disrupted molecules. The cosmic rays and the SR^{90}, however, are quite similar even in this respect.

The reader will recall that the spacing of the ionization depends only on the charge and the speed of the ionizing particle. The ionizing particle from the Sr^{90} is an energetic beta ray, which has a charge of one and a speed close to that of light. A large part of the background radiation which reaches our bones comes from the cosmic rays. The main portion of the cosmic rays is due to the mesons. The meson, like the beta ray, has a unit charge and a speed close to that of light. The two particles may therefore be expected to produce identical biological effects. The only difference between their effects is that the beta ray does not have enough energy to leave the bones, while the meson is so energetic that it deposits its energy both in our bones and throughout our whole body. Thus if we compare a Sr^{90} dose with the same dose of cosmic rays the same effect to the bones must be expected. But the cosmic rays give rise to additional effects in our bodies.

The total background dose to the bones is about 0.15 roentgens per year for the average person living at sea level in the United States. Of this amount, about 0.035 roentgens is due to cosmic rays. At higher altitudes the cosmic ray dosage increases. In Denver, at an altitude of 5000 feet, the cosmic rays contribute 0.05 roentgens per year.

The above numbers should be compared with the present level of worldwide fallout radiation to the bones: about 0.003 roentgens per year (from Sr^{90} and other sources). The fallout radiation is thus only a few per cent of the natural cosmic radiation. It is small even when compared to the variation of cosmic ray intensity between sea level and 5000 feet.

A correlation between the frequency of leukemia and bone cancer, and the intensity of natural radiation has been looked for. Some statistics for the year 1947, before weapons testing began, are available. They show the number of cases of these diseases occurring in that year per 100,000 population.

	Bone Cancer	*Leukemia*
Denver	2.4	6.4
New Orleans	2.8	6.9
San Francisco	2.9	10.3

The extra radiation that one gets in Denver from cosmic rays is many times greater than the fallout radiation. But the table shows no increased incidence of bone cancer or leukemia. On the contrary—the incidence of these diseases is actually lower in Denver.

Not all of the natural background radiation is due to cosmic rays. Part of the background comes from natural radioactive elements in the soil and in the drinking water. These include uranium, $potassium^{40}$, and thorium and radium. Radium behaves like calcium and strontium, and get deposited in our bones. All these effects are, to the best of our knowledge, at least as intensive in the Denver area as in San Francisco or New Orleans.

One possible explanation for the lower incidence of bone cancer and leukemia in Denver is that disruptive processes like radiation are not necessarily harmful in small enough doses. Cell deterioration and regrowth go on all the time in living creatures. A slight acceleration of these processes could conceivably be beneficial to the organism. One should not forget that while radiation can cause cancer, it has been used in massive doses to retard and sometimes even to cure cancer. The reason is that some cancer cells are more strongly damaged by radiation than the normal cells.

In spite of the table, however, there may actually be an increased tendency toward bone cancer and leukemia that results from living in Denver. If so—and this is the main point —the effect is too small to be noticed compared to other effects. We must remember that Denver differs from New Orleans and San Francisco in many ways (besides altitude), and these differences may also influence the statistics.

A more thorough consideration of the background radiation gives further evidence that this radiation is more important than the present or expected effects of Sr^{90}. The radium deposited in

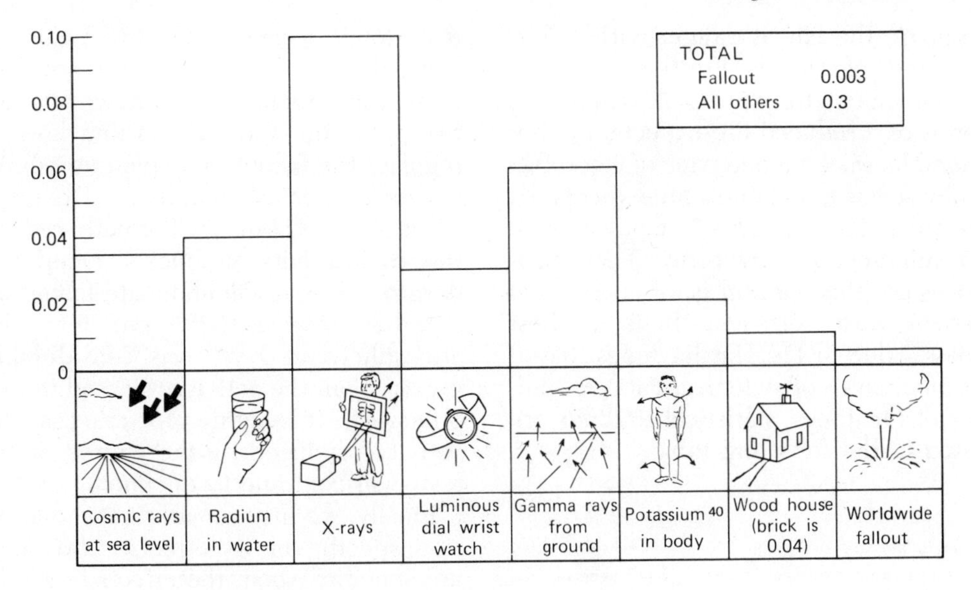

How people get radiation (average dose in roentgens per year).

our bones from drinking water has been observed to reach values as high as 0.55 roentgens per year. Furthermore, the heavier and slower alpha particles emitted by radium cause ionization processes which occur in closer spacing and are therefore more damaging than the ionization due to Sr^{90}. To make things worse radium is deposited in our bones in little nodules (hot spots). Thus the possibility of local damage is enhanced.

The background radiation to which we are exposed varies for some unexpected reasons. It has been pointed out recently that brick may contain more natural radioactivity than wood. The difference between living in a brick house and living in a wood house could give rise to ten times as much radiation as we are currently getting from fallout. (The additional radiation from the brick might be as much as 0.03 roentgens per year.)

Human beings are subject to radiation not only from natural sources, but also from man-made sources. One of these is wearing a wrist watch with a luminous dial. Another is having X-rays for medical purposes. Both of these sources give much more radiation than the fallout.

Of all ionizing radiation to which we are exposed the X-rays are most important. In some cases medical X-rays have intensities which are noticeably harmful. Yet this damage is practically always of little consequence compared to the advantage from correct recognition of any trouble that the X-ray discloses.

We may summarize in this way. Our knowledge of the effects from the fallout is deficient. We cannot say exactly how many lives may be impaired or shortened. On the other hand, our knowledge is sufficient to state that the fallout effect is below the statistically observable limit. It is also considerably less than the effect produced by moving from sea level to an elevated location like Denver, where cosmic radiation has a greater intensity. It is also less than having a chest X-ray every year. In other words, we know enough to state positively that the danger from the world-wide fallout is less than many other radiation effects which have not worried people and do not worry them now.

We have compared radiation from the fallout with radiation from other sources. It is also possible and helpful to compare the fallout danger with different kinds of dangers. For this purpose it is convenient to express all dangers in terms of a reduced life-expectancy. For example, smoking one pack of cigarettes a day seems to cut one's life-expectancy by about 9 years. This is equivalent to 15 minutes per cigarette. That cigarettes are this harmful is, of course, not known with certainty. It is a "best guess," due to Dr. Hardin Jones, based on an analysis of statistical data. A number of Dr. Jones' statistical findings are listed in the following table:[1]

	Reduced Life Expectancy
Being 10 per cent overweight	1.5 years
Smoking one pack of cigarettes a day	9 years
Living in the city instead of the country	5 years
Remaining unmarried	5 years
Having a sedentary job instead of one involving exercise	5 years
Being of the male sex	3 years
Automobile accidents	1 year
One roentgen of radiation	5 to 10 days
The world-wide fallout (lifetime dose at present level)	1 to 2 days

The reader will see that the world-wide fallout is as dangerous as being an ounce overweight or smoking one cigarette every two months.

The objection may be raised that the fallout, while not yet dangerous, may become so as more nations develop and test atomic weapons. On this point we can only say that the future is not easy to predict. Some factors, however, justify optimism. We are learning how to regulate the fallout by exploding bombs under proper surroundings. Development of clean bombs will greatly reduce the radioactivity produced. Deep underground tests will eliminate fallout altogether. The activity put into the atmosphere in 1954 was considerably greater than the activity released in any other year. It is highly probable that the activity produced by the United States tests will continue to decline.

Finally, we may remark that radiation is unspecific in its effects. Chemicals are specific. About the effects of a new ingredient in our diet, in our medicine, or in the air we breathe, we know much less than we know about radiation. If we should worry about our ignorance concerning our chemical surroundings as we worry about the possible effects of radiation, we would be condemned to a conservatism that would stop all change and stifle all progress. Such conservatism would be more immobile than the empire of the Pharaohs.

It has been claimed that it is wrong to endanger any human life. Is it not more realistic and in fact more in keeping with the ideals of humanitarianism to strive toward a better life for all mankind?

[1]The last line of the table is based on our own estimates.

8.

Chemical and Biological Warfare

Seymour M. Hersh

Mr. Hersh has covered the Pentagon for United Press International and he was press secretary for Senator Eugene McCarthy during his campaign for the Democratic presidential nomination.

In his book *Chemical and Biological Warfare*, Mr. Hersh offers a comprehensive and thoroughly documented investigation of CBW (chemical and biological warfare). He has lifted the veil of secrecy from this controversy-charged, closely guarded subject to allow the general public scrutiny for the first time.

It is rarely mentioned, since nuclear weapons dominate the scene, that there is a biological agent so powerful that a mere 8 ounces properly dispersed would be enough to wipe out the entire population of the world.

Table of Chemical and Biological Agents

The Chemical Agents

NERVE GASES

GB: An odorless, colorless, volatile gas that can kill in minutes in dosages of 1 milligram, approximately 1/50 of a drop. In the U.S. arsenal since the late 1940's, it is also known as Sarin. The gas kills by paralyzing the nervous system.

VX: Another odorless gas that, unlike GB, does not evaporate rapidly or freeze at normal temperatures. Because of its low volatility, it is effective for a longer period of time. VX also is capable of killing in 1 milligram doses and, like GB, paralyzes the nervous system in minutes.

INCAPACITATING AGENTS

BZ: A gas that is either a psychochemical or a strong anesthetic which can produce temporary paralysis, blindness, or deafness in its victims. BZ has also been known to cause maniacal behavior. Its precise makeup is secret.

RIOT CONTROL GASES

CN: A non-lethal gas with a deceptive, fragrant odor similar to apple blossoms. The agent, now in use in Vietnam, is a fast-acting tear gas that also acts as an irritant to the upper respiratory system.

CS: An improved, more toxic tear gas that quickly causes tearing, coughing, breathing difficulty, and chest tightness. Can temporarily incapacitate men in twenty seconds. Heavy concentrations cause nausea. It is now used in Vietnam.

HARASSING AGENTS

DM: A pepper-like arsenical gas that causes headaches, nausea, vomiting, chest pains for up to two or three hours. It can be lethal in heavy doses and has been blamed for some deaths since its first use in Vietnam in 1964. DM is widely known as adamsite and was used in World War I.

HD: A pale yellow gas with the odor of garlic, popularly known as mustard gas. Causes severe burns to eyes and lungs and blisters skin after exposure, but onset of symptoms is delayed from four to six hours. Can kill in heavy concentrations. Mustard, like VX, is not volatile and is usually effective for days after its use. It caused one-fourth of the U.S. gas casualties in World War I.

DEFOLIANTS AND HERBICIDES

2,4-D: A weed-killing compound known as dichlorophenoxyacetic acid that has relatively short persistence in the soil and a relatively low level of toxicity to man, if properly dispersed. Heavier concentrations can cause eye irritations and stomach upsets, however. Dangerous to inhale. Usually used in Vietnam along with 2,4,5-T (trichlorophenoxyametic acid), which has similar—although somewhat more toxic—properties. Effective against heavy jungle.

Cacodylic Acid: An arsenic-base compound used against rice plants and tall grass. Strong plant killer that gives quick results. One serious restriction on its use is the possibility that heavy concentrations will cause arsenical poisoning in humans. Widely used in Vietnam. It is composed of 54.29 per cent arsenic.

Biological Agents

Anthrax: An acute bacterial disease that is usually fatal if untreated when it attacks the lungs (pulmonary anthrax). Death can result in twenty-four hours. Found naturally in animals, which must be buried or burned to prevent contamination. Symptoms include high fever, hard breathing, and collapse. Also known as woolsorters' disease.

Brucellosis: Bacterial disease usually found in cattle, goats, and pigs. Marked by high fever and chills in humans. Also known as undulant fever. Fatal in up to 5 per cent of untreated cases. Symptoms can linger for months.

Encephalomyelitis: Highly infectious viral disease that appears in many forms and gradations: it can be simply debilitating or fatal. Venezuelan equine encephalomyelitis (VEE) kills less than 1 per cent of its victims and lasts as few as three days; Eastern equine encephalomyelitis (EEE) is fatal about 5 per cent of the time, if untreated, and can seriously cripple the central nervous system of survivors.

Plague: Acute, usually fatal, highly infectious bacterial disease of wild rodents found in two forms—bubonic and pneumonic. Symptoms of bubonic plague include small hemorrhages, and the black spots that led the disease to be commonly known as the "black death" during the massive epidemics of the past. Pneumonic plague is highly infectious because it is spread from man to man via coughing. Symptoms include fever, chills, rapid pulse and breathing, mental dullness, coated tongue, and red eyes.

Psittacosis: Viral infection in birds that is transmissible to man, with symptoms of high fever, muscle ache, and disorientation. Disease can be mild, and last less than a week, or can cause death in upwards of 40 per cent of those afflicted. Complete convalescence may take months.

Q-fever: Acute, rarely fatal rickettsial disease usually found in ticks, but also found in cattle, sheep, goats, and some wild animals. The Q-fever organism can remain alive and infectious in dry areas for years. Rarely fatal but the resulting fever may last up to three months.

Rift Valley Fever: Viral infection of sheep, cattle, and other animals that can be transmitted to humans, usually to the male. Symptoms include nausea, chills, headaches, and pains, but the disease is mild: despite the severity of symptoms deaths are rare and acute discomfort lasts only a few days. Also believed to be more virulent among Asians.

Rocky Mountain Spotted Fever: An acute rickettsial disease transmitted to man by the tick. One of the most severe of all infectious diseases. Can kill within three days. Fevers range up to 105 degrees F. Often found in northwestern United States, but susceptibility to the disease is general. Highly responsive to treatment.

Tularemia: A bacterial disease marked by high fever, chills, pains, and weakness. Acute period can last two to three weeks. Sometimes causes ulcers in mouth or eyes, which multiply. Untreated, its mortality rate is between 5 and 8 per cent. Highly infectious, and usually found in animals, fowls, and ticks. Also known as rabbit fever.

The History of Chemical and Biological Warfare

"The annals of history," notes a 1960 Army Chemical Corps handbook on chemical and biological warfare (CBW), "show that down through the ages man has sought to enlist the aid of chemistry and disease in his conduct of warfare, but it was not until the 20th century that science made it possible."

Historians can argue otherwise: chemicals and incapacitating smoke screens have been successfully used by combatants for thousands of years. The 20th century improvements have merely escalated the age-old practices.

In ancient days, as today, the use of germs and gases was viewed as an especially abhorrent maneuver or threat. The wars of ancient India in about 2000 B.C. were fought with smoke screens, incendiary devices, and toxic fumes that caused "slumber and yawning." Arsenical smokes were understood during the Sung Dynasty. Thucydides tells of the use of gas during the seige of Plataea in 429 B.C., during the Peloponnesian War. The Spartans saturated wood with pitch and sulphur, placed it under the city walls, and set fire to it. "The consequence was a fire greater than any one had ever yet seen produced by human agency," the Greek historian wrote. Choking, poisonous fumes rose, but a sudden rainstorm is said to have put out the fire and averted the danger. Similar tactics were successful in 424 B.C. during the siege of Delium.

In the Middle Ages chemical warfare was put to similar use. In 1456 Christian Belgrade was saved from attacking Turks by an alchemist who prepared a poisonous mixture. The Christians dipped rags in the chemical and burned them, creating a toxic cloud.

In 1855 a British naval officer, Admiral Lord Dundonald, proposed using sulphur to help defeat the Russians during the Crimean War. The English government, after a long study, concluded that the effects of the chemical would be so horrible that no honorable combatant could use it. Near the end of the U.S. Civil War, a New Yorker named John W. Doughty wrote to Secretary of War Edwin M. Stanton and suggested the manufacture of a liquid chlorine gas shell. At the end of the 19th century, during the Boer War, English troops used picric acid in artillery shells. Once on the ground, the shells released an explosive gas known as lyddite. The Boers protested to the English but the issue quickly became academic: the shells were not very effective.

World War I

Despite a major international disarmament agreement on the use of gas shells between the Boer War and the outbreak of World War I,[1] both sides made steady use of phosgene, chlorine, and mustard gas, along with many less toxic agents. About 17,000 chemical troops were employed by the Allies and their enemies and 1.3 million casualties, including 91,000 deaths, were attributed to gas warfare. U.S. forces were not involved in gas attacks until February 25, 1918, when they were hit by German phosgene shells.[2] The first offensive use of gas by the U.S. was against the Germans in June, 1918.

Although mustard gas was not used until late in the war, more than 9 million shells filled with it were fired by both sides, causing an estimated 400,000 casualties. A total of 124,200 tons of toxic gases were used in the war, with the Allies firing about 58,000 tons. U.S. troops, which did not enter the war until 1917, utilized only about 1,100 tons of gas.

Most military historians date the use of gas in World War I from April 22, 1915, the day the Germans released a lethal cloud of chlorine gas on the French lines at Ypres. Although a German prisoner had forewarned of the attack—he later was known as "the traitor of Ypres"—the French were taken unawares by the new gas. More than 5,000 soldiers were killed and another 10,000 injured as the gas caused massive confusion and panic, setting the army in total retreat, leaving a four-mile gap in the front line, and opening the road to the English Channel. The Germans failed to follow up their advantage; apparently they had not anticipated such success.

The attack at Ypres was not, however, the first gas attack of the war. The French had initiated the use of hand and rifle grenades filled with tear gas in 1914 and the Germans retaliated with tear gas artillery shells. The gas was not lethal and neither the Allies nor the Germans provided their troops with masks for protection.

The attack at Ypres was the only significant gas battle of the war. By late April, 1915, the Allies had rushed improvised gas masks to the front lines and on May 1 a German infantry attack preceded by a chlorine cloud was stopped cold. Defensive measures eventually were so successful that gas lost its potential as a strategic weapon in World War I, although by the end of the war the United States had developed lewisite, a new type of blister gas that could quickly blister the skin and penetrate the body.

Summing up the developments in World War I, one military writer has noted that "gas gave the side first using it a temporary advantage only. A new development by one side was in time matched by the other. Gas could inflict casualties, but to no greater extent that those inflicted in turn by an enemy who was prepared to retaliate." Despite this, the history of the use of gas during the war is one of steady increases in lethality. Chlorine replaced tear gas, phosgene replaced chlorine, and mustard gas eventually replaced chlorine.

Worldwide revulsion over the use of such weapons, with the United States in

[1]Peace conferences were held at the Hague in 1899 and 1907, during which the major nations approved, over U.S. objections, a resolution outlawing "the use of projectiles the sole object of which is the diffusion of asphyxiating or deleterious gases." Twenty-six nations signed the treaty in 1899; the United States refused and Great Britain delayed until 1907. Historians attribute the treaty's lack of effectiveness to its narrow scope; "projectiles" did not take into account smoke pots and similar disseminating devices that were used to spread gases during World War I.

[2]Phosene is a choking gas that was used heavily in World War I. It attacks the respiratory tract and, in extreme cases, can cause swelling of membranes and death from lack of oxygen.

the forefront, led to the convening of a Geneva Conference in 1925, which outlawed all use of asphyxiating, poisonous, or other gases. The United States signed the treaty but the Senate Foreign Relations Committee refused to ratify it in 1926 after a rare closed-door debate. The Geneva protocol, which eventually was ratified by thirty-two nations, was openly violated only once before World War II, when Italy used mustard gas against unprepared Ethiopians in the Abyssinian campaign of 1936. The attack was widely condemned in the United States and elsewhere.

World War II

By the 1930's, U.S. intelligence was aware that Russia, Japan, and Germany were carrying out active research programs in chemical and biological warware. Japan, which did not sign the 1925 Geneva agreement, was accused on a number of occasions of employing gas in its war against China.

Accusations of chemical warfare were few during World War II among the major powers and gases were apparently never authorized for combat use, although Germany had developed a new superclass of chemical killers—the nerve gases.[3] Odorless and sightless, these agents are seventy-five times more lethal than the mustard gases that caused most of the gas casualties in World War I. Their discovery came by accident: While investigating insecticides in 1936, Dr. Gerhard Schrader, a research chemist with I. G. Farbenindustrie in Leverkusen, Germany, synthesized a complex organophosphorous ester which came to be known and manufactured under the name tabun.[4] By 1938 similar research, carried on under stringent secrecy, led to the development of a far more toxic agent, known as sarin. A third nerve gas, soman, was developed in 1944. All can kill in minutes.

The nerve gases are similar in chemical make-up to the insecticides and pesticides now in common use.[5] By April, 1945, Germany had manufactured 12,000 tons of tabun at a secret plant near the Polish border that was opened in 1942. (Besides nerve agents, the Germans had manufactured vast stores of more conventional agents. In 1946, at a Washington, D.C., Medical Society banquet, a Chemical Corps general revealed that the Allies had found 250,000 tons of gas stored at St. Georgian in Austria in the last days of the war.)

[3]Gas warfare may not have been used in World War II, but it did claim more than 600 victims. Eighty-three sailors were killed and 534 seriously injured when German aircraft bombed a U.S. cargo ship loaded with approximately 100 tons of 100-pound mustard bombs in early December, 1943. The ship—the SS "John Harvey"—was berthed in the harbor at Bari, Italy; its cargo was being sent to the Allied command in Europe to be used in retaliation in case the Germans initiated gas attacks. The German raid also destroyed sixteen Allied cargo ships stacked up in the harbor waiting to be unloaded. In the confusion, harbor officials forgot that the "Harvey" was carrying mustard. The gas, loaded into the shells in liquid form, spread across the oil-covered waters where sailors were struggling to get rescued. None of the victims knew he had been exposed to the agent until hours later, although many recalled that they had noted "an odor of garlic." Fourteen hours after the raid, port authorities discovered broken mustard shell casings and informed the hospital; by then it was too late. The first casualty died four hours later. Many of the deaths were linked to severe chemical burns and internal lung damage due to inhaling the intense mustard fumes. The incident was classified top secret until 1959.

[4]I. G. Farbenindustrie also developed Zyklon B, the lethal gas used by the Nazis to kill Jewish concentration camp prisoners during World War II.

[5]Strong doses of the more potent insecticides, such as parathion, can kill men as well as mosquitoes. Two tragic incidents involving the chemical took place within months of each other in 1967. In September seventeen children were killed and hundreds were made violently ill after eating food mixed with parathion in Tijuana, Mexico. Eighty persons, only four above 17 years of age, were killed in Chiquinquira, Columbia, in November after eating bread contaminated by parathion. The *New York Times* reported that less than a pint of the insecticide caused the deaths. More than 600 persons were affected at the height of the outbreak.

The existence of the German nerve agents was not known to the Allies until some tabun chemical shells were captured and analyzed in 1945. For the sake of secrecy during the war, the Germans had called its nerve agent compound trilon, the name of a popular soap manufactured in Germany. After the war the Soviet Union, Great Britain, and the United States immediately began their own research. It is unclear why the Germans did not use the nerve gases in combat, but it is known that Hitler sent the newly developed agents to his forces in the field. Albert Speer, Hitler's Minister of Production, told the Nuremburg court in 1947 that the Nazi leader had apparently intended that the new gases be used.

"I was not able to make out from my own direct observations whether gas warfare was to be started, but I knew from various associates . . . that they were discussing the question of using our two new combat gases: tabun and sarin," Speer testified. "They believed that these gases would be of particular efficacy and they did in fact produce the most frightful results . . . for the manufacture of this gas, we had about three factories, all of which were undamaged and which until November, 1944, were working at full speed. When rumors reached us that gas might be used, I stopped its production in November, 1944." Speer explained that "all sensible army people turned gas warfare down as being utterly insane, since, in view of [America's] superiority in the air, it would not be long before it would bring the most terrible catastrophe upon German cities. . . . " By the time Speer testified, 1,000 tons of German-made tabun and sarin had already been secretly shipped to the United States, labeled "chlorine."

A few months after the war, Brigadier General Charles E. Loucks, head of U.S. chemical warfare operations in the Pacific Theater, declared that the Japanese had used poison gas against American troops "in a few" isolated instances, notably in New Guinea late in the war. No Americans were killed in the attacks, the General said. He added that Japanese officials had told him that they did not know of the use of gas, but conceded that there might have been some isolated battlefield attacks with it.

According to Loucks, the Japanese had an efficient chemical warfare school at Narashino, but abandoned the manufacture of agents after the United States pledged in 1943 that it would use gas only in retaliation. He added that the Japanese had studied more than 1,000 compounds during the war looking for more lethal agents, but found none as effective as the standard gases used in World War I.

Many of the details of the World War II improvements in chemical warfare, such as Germany's discovery of the nerve agents, were made public by the U.S. Army after the war, but information about research into biological warfare—both during and since the war—has been blanketed in rarely yielding secrecy. Nazi Germany was known to have begun large-scale biological research in 1936. By 1938, Russia, a repeated target of German threats, was warning publicly that "if our enemies use such [biological] methods against us, I tell you we are prepared, and fully prepared, to use them also, and to use them against aggressors on their own soil."

In 1940 Britain set up its biological warfare research center at the Ministry of Supply Station in Porton. Canada similarly established its research center in Suffield along the Trans-Canada Highway. In the fall of 1941 the United States moved to follow suit. The War Department asked the National Academy of Sciences to appoint a committee to survey the current biological warfare (BW) situation and its future possibilities. The committee—known as the WBC Committee—concluded that germ warfare was distinctly feasible and in the summer of 1942 the War Research Service was established, with George W. Merck as Director. Camp Detrick was opened a year later in stringent secrecy.

On January 3, 1946, the War Department made public a special report on biological warfare telling, for the first time, some of the factors that had led to the full-time utilization of the germ warfare center in 1944.

"In December 1943," the study, widely known as the Merck Report,[6] said, "the Office of Strategic Services reported to the Joint Chiefs of Staff that there were some indications that the Germans might be planning to use biological warfare agents. While the evidence that the Germans might use such agents was inconclusive, there was considerable concrete information available from work which had been carried on in the United States, the United Kingdom, and Canada that attack by biological warfare was feasible. Accordingly, it was decided in January, 1944, to step up all work in this field, particularly in terms of the protection of troops against possible enemy use of those weapons, and to transfer a large part of the responsibility for the biological warfare program to the War Department."

It later was learned, the Merck Report said, that the United States was ahead of the Nazis in the development of germ warfare, but little information on Japanese progress was available until well after the end of the war.

Most of the Allied biological warfare concern was over Germany. Major General Brock Chisolm, who headed the Canadian medical effort during the war, revealed in a private speech in 1957 that Allied intelligence first feared, upon hearing of the German V-1 launching sites in northern France and Belgium, that the Germans were planning to fill them with biological warheads, particularly botulinus toxin, eight ounces of which, properly dispersed, could kill the world's population. Chisolm reported that his government sent over 235,000 doses of an antidote to London, and self-inoculating syringes were given to 117,-500 British, American, and Canadian troops. He added:

"The information that this had been done was carefully fed into German intelligence sources, so that they knew that we were ready for biological warfare. When the V-1 attack was launched in June, 1944, and the first flying bomb went off with a big bang, showing that it only contained normal high explosives, the general staffs all heaved an immense sigh of relief."

The Allied troops carried their antidote along on the invasion of Normandy.

Biological Warfare—The Accusations

After the war, it was learned that Germany—although far ahead of the Allies in chemical preparations—had done relatively little with biological warfare. Yet in 1946 a Russian war crimes tribunal charged that the German High Command had prepared plans to wage germ warfare and supported the charges with a sworn affidavit from Major General Walter Schreiber, former German army medical specialist who was then a Russian prisoner. Such charges do not appear in any U.S. versions of Nazi germ warfare development, although Schreiber was eventually released by the Russians and came to the United States in 1951. Ac-

[6]Although newspaper reports in the late 1940's consistently referred to the Merck Report as being "suppressed," it was reprinted in the March, 1946, *Military Surgeon* and also in the October, 1946, issue of the *Bulletin of the Atomic Scientists.* Apparently the Pentagon stopped making copies of the report available to newsmen after a secrecy edict about such research was issued later in 1946. Merck was a special consultant to the War Department for biological warfare.

cording to State Department files, the Army Air Force sponsored his trip and put him to work in the School of Aviation at Randolph Field, Texas. Jewish veterans' groups protested the Army's action, and in 1952 the former Nazi general left the United States for Argentina. It is not known if Schreiber did any research into CBW while in the United States.

U.S. disclosures on the Japanese development of germ warfare are also much less explicit than the Russian reports. On Christmas Eve, 1949, Moscow radio announced that twelve Japanese Army officials had been indicted on charges of causing epidemics in China early in the war by spreading bubonic plague, cholera, typhoid, typhus, and anthrax germs. The Russians claimed that only their swift victory over Japan's Kwantung Army in Manchuria prevented the Japanese from using germ warfare against the Allies. They said Japan had set up secret laboratories in Manchuria and was producing bacteria in vast quantities.

Over the next few days the charges were elaborated. On December 27, Moscow radio said Detachment 731 of the Kwantung Army—the biological warfare unit—had used American war prisoners as guinea pigs in a number of experiments. On December 28 they said one of the defendants had confessed that the germ warfare detachment had been set up in 1936 by orders of Emperor Hirohito and Minister of War Hideki Tojo. On December 29 *Pravda* correctly reported that the United States was actively engaged in biological warfare research and also accused America of deliberately protecting a number of Japanese war criminals. The Russian military newspaper *Red Star* later charged that eighteen Japanese were working on biological warfare in U.S. labs in Japan. The twelve Japanese prisoners were eventually sentenced to jail in terms varying from two to twenty-five years.

The Russian charges attracted wide press attention throughout America. The U.S. response was to refute some of them and ignore others. Joseph B. Keenan, prosecutor at the Tokyo war crimes trials, said an on-the-spot investigation by American and Chinese officials thoroughly "exploded" claims that the Japanese had dropped germs on China during the war. U.S. officials did not deny, however, that there was widespread evidence of a massive Japanese germ warfare effort. In 1946, Hanson Baldwin, the *New York Times* military writer, reported that "the Japanese are known to have experimented considerably with biological warfare. . . . The Japanese had developed, before the war ended, a crude anthrax bomb, and they had a BW factory near Harbin, Manchuria, which was producing toxins or bacteriological poisons. The Russians have taken over this factory and are presumed to have removed its equipment to Russia."

And in a special note issued to newsmen before release of the Merck Report in 1946, the War Department said that late intelligence reports "show that Japan had made definite progress in biological warfare. . . . Intensive efforts were expended by Japanese military men toward forging biological agents into practical weapons of offensive warfare." The note added:

"Modifications of various weapons developed through research in their laboratories were field-tested at Army Proving Grounds where field experiments were also conducted in the use of bacteria for purposes of sabotage. . . . While definite progress was made, the Japanese had not at the time the war ended reached a position whereby these offensive projects could have been placed in operational use."

The War Department concluded by reporting there was "no evidence that the enemy ever resorted to this means of warfare."

The Russian charges and trial were left as just another forgotten propaganda incident when the Korean War broke out in 1950. But in 1955 the Tokyo magazine *Bungei Shunju* published an eyewitness account of Japanese germ warfare tests during World War II in which 1,500 to 2,000 human guinea pigs allegedly died. The story, written by Hiroshi Akiyama, said the germ warfare center was near Harbin and had been masked as a Red Cross unit; it had been hurriedly destroyed in August, 1945, when Russia entered the war. The magazine was described at the time by U.S. government officials as having a good reputation with no pro-Communist connections. "I hereby dare to make public this report," Akiyama wrote, "because I wish to help prevent the third world war." He said the guinea pigs "were mostly Chinese and Manchurians and occasionally Koreans, Japanese, Russians, and halfbreeds of uncertain nationalities." Akiyama said he was a drafted civilian assigned to a station where humans were injected daily with cholera, typhus, or bubonic plague bacteria. When the orders came to destroy the center, Akiyama wrote, the prisoners still held were killed in an orgy of poison and machine-gun bullets. All prisoners were given poisoned food for breakfast. Those who escaped were shot down. The bodies later were burned and their ashes scattered.

His story coincides in many ways with testimony presented at the trial of the twelve Japanese. The Foreign Languages Publishing House in Moscow published *Materials on the Trial* in 1950 and distributed copies in English in the United States. Witnesses told of two germ warfare units in Manchuria, Detachments 731 and 100. According to the documents supplied in the English version

"These detachments were composed of expert bacteriologists and their extensive research and technical personnel was directed by some of the leading bacteriologists of Japan. The scale of the work conducted by the bacteriological detachments is indicated, among other things, by the fact that Detachment 731 alone had a personnel total of about 3,000."

Detachment 731 was located about twenty kilometers from Harbin.

According to the Russians, the detachment had eight divisions geared to research on breeding, testing, dissemination, training, and the mass production of germs, including large autoclaves and cooling chambers for cultivation and storage. The trial record estimated that "Detachment 731 alone was capable of breeding, in the course of one production cycle, lasting only a few days, no less than 30,000,000 billion microbes. . . . Detachment 731 and its branches also engaged in the wholesale breeding of fleas for their infection with germs," and possessed 4,500 incubators for breeding fleas on rats and mice.

The witnesses at the Moscow trial told of gruesome experiments. Major Karasaw Tomio, chief of a section of Detachment 731, was reported to have testified that he

". . . personally was present on two occasions at the Anta proving ground when the action of bacteria was tested on human beings under field conditions. The first time I was there [was] toward the end of 1943. Some ten persons were brought to the proving ground, were tied to stakes, which had been previously driven into the ground five meters apart, and a fragmentation bomb was exploded by electric current fifty meters away from them. A number of the experimentees were injured by bomb splinters and simultaneously, as I afterwards learned, infected with anthrax, since the bomb was charged with these bacteria. . . .

The second time I visited the proving ground was in the spring of 1944; about ten people were brought there and, as on the first occasion, tied to stakes. A cylinder filled with plague germs was then exploded at a distance of roughly ten meters from the experimentees."

Another witness, Lieutenant Colonel Nishi Toshihide of Detachment 731, testified:

"In January, 1945, in my presence, Lt. Col. Ikari, Chief of the 2nd Division of Detachment 731, and Futaki, a research official of this division, performed an experiment at the detachment's proving ground near Anta Station infecting ten Chinese war prisoners with gas gangrene. The ten Chinese prisoners were tied to stakes from ten to twenty meters apart, and a bomb was then exploded by electricity. All ten were injured by shrapnel contaminated with gas gangrene germs, and within a week they all died in severe torment."

Other witnesses testified that up to 600 prisoners passed through Detachment 731 annually, but could not say how many were killed.

If the U.S. occupiers of Japan ever found any corroborating evidence, it was shunted aside, although no specific denials were apparently ever made. In 1965 George Bilainkin, a British writer, published *Destination Tokyo,* in which he said President Harry S. Truman had ordered the 1945 atomic bombing of Japan because he feared that nation was ready to unleash germ warfare. The book said a Japanese general had told the Russian court in 1950 that the Japanese research program's aim was "to employ the bacteriological weapon against any other enemy state or enemy army, the United States and Britain in particular." Truman did not comment on the report.

In February, 1952, the Communist world made a new charge of germ warfare—this time the United States was accused of dropping germ bombs on North Korea. The North Korean and Communist Chinese charges were based on testimony said to have been volunteered by more than thirty captured U.S. Air Force officers, some captured intelligence agents, and others who said they had discovered vast quantities of fleas and other pests or insects shortly after American planes had flown over the area. Eventually Communist China published six pilots' confessions and photographs purporting to show "American germ bombs" and the diseased flies they allegedly carried into North Korea. The United States repeatedly and emphatically denied the charges and accused the Communists of brainwashing the captured pilots.

Communist China immediately moved to support its charges by setting up "The International Scientific Commission for the Investigation of the Facts Concerning Bacterial Warfare in Korea and China," a group that included scientists from Sweden, France, Italy, Russia, Brazil, and England. The English representative was Dr. Joseph Needham, a renowned scientist who had served as counselor to the English embassy in Chunking and later was director of the Department of Natural Sciences of the United Nations Educational, Scientific, and Cultural Organization (UNESCO). The commission concluded, after a long investigation, that "the peoples of Korea and China did actually serve as targets for bacteriological weapons. These weapons were used by detachments of the armed forces of the U.S.A., who used for this many and various methods; some of these are a continuation of methods used by the Japanese army in the second world war." It filed a lengthy report with the United Nations October 8, 1952. The 700-page study, made available to U.S. libraries, cited the use in North Korea and China of cholera-infected clams, anthrax-infected feathers, plague- and yellow-fever-infected

ing from Cuban exiles in the United States. The emigrants had told Miami newspapers the year before that Cuba was preparing a germ warfare attack against the United States in a secret laboratory at Soroa, in Pinar Del Rio Province. This charge was made by Oscar Alcalde Ledon, described as a former director of the Cuban Academy of Science who had escaped from Cuba in mid-1963 with twenty other refugees aboard a small boat. Alcalde said some Cuban officials had suggested to him that "it now is very easy for the Cuban government to introduce foot and mouth disease into the United States. . . ."

America First: CBW Policy Today

On November 21, 1964, the *New York Times* carried a front-page story describing the booming U.S. research efforts in chemical and biological warfare (CBW). The dispatch noted: "It is a long-established United States policy that chemical and biological agents will never be used by American forces except in retaliation for a chemical-biological attack."

The statement was based on a 1943 speech by Franklin D. Roosevelt that has been permitted to stand since then as the clearest statement of U.S. CBW policy. Roosevelt, after noting reports that the Axis powers were contemplating the use of poison gases, warned that "use of such weapons has been outlawed by the general opinion of civilized mankind. This country has not used them, and I hope that we never will be compelled to use them. I state categorically that we shall under no circumstances resort to the use of such weapons unless they are first used by our enemies."

The 1960 Senate disarmament study on CBW described the Roosevelt speech as "the last official statement of U.S. policy."

In fact, by the mid-1950's official military policy toward the use of CBW agents had undergone a dramatic change, one that still remains secret. One hint of the change came during 1958 Congressional hearings when Major General William M. Creasy, Director and representative of the Army Chemical Corps, and Representative Gerald R. Ford, Jr., Michigan Republican, had this heavily censored exchange:

Creasy: "First, I will start off with the national policy." (Discussion off the record.)

Ford: "May I ask how long that policy has been in effect?"

Creasy: "Since about October 1956, about a year and a half ago. The national policy has been implemented by a Department of Defense directive."

The change in policy was simple enough: After 1956 the military was free to wage chemical and biological warfare on a first-strike basis during conventional warfare.

The policy change came shortly after a high-level civilian advisory committee recommended to Secretary of the Army Wilber M. Brucker that CBW agents be developed for "actual use" if necessary. The civilian committee, headed by Otto N. Miller, Vice-President of the Standard Oil Company of California, decried the public conception that CBW is "horrifying in character" and said the public should be informed that such weapons have a proper place in military planning.

Evidence of the new military directives is found in Army manuals dating back to 1954. In that year, Field Manual 27-10, *Law of Land Warfare,* contained this provision: "Gas warfare and bacteriological warfare are employed by the United States against enemy personnel only in retaliation for their use by the enemy." By 1956, that provision had disappeared from subsequent editions lice, fleas, mosquitoes, rodents, rabbits, and other small animals as well as such non-living things as toilet paper, envelopes, and fountain pens filled with germ-laden ink. The study included exhaustive fold-out charts depicting specific incidents and findings and photographs of animals said to have been infected, the bacteria used to infect them, and the bombs used to disseminate the bacteria. The report has received scant recent attention since, but in 1966 it was described by Robin Clarke, editor of *Science Journal*, published in London, as "such a curious mixture of blatant propaganda and what appears to be accurate observation that it is difficult to know how much credence should be attached to it."

In April, 1953, the United States asked the General Assembly to place on its agenda an item calling for an impartial investigation by the United Nations. But Communist China and North Korea refused to grant access to their territories, and the issue died a few months later when the Korean armistice was signed and the American fliers were released.

The final chapter of the Korean germ warfare accusations took place in 1958 when the United States pressed sedition charges against three Americans accused of publishing the false germ war charges in a Shanghai-based magazine they edited. In a year-long battle, the Eisenhower administration failed to get convictions against John W. Powell, editor of the *China Monthly Review*, his wife, Sylvia, and an associate, Julian Schuman of New York, but the case did bring out more conflicting evidence about the Communist charges.

In January, 1959, the government acknowledged it had had the weapons to wage germ warfare in 1952 but said they were not sent overseas. A sworn statement said that the "bacteriological warfare capability [of the United States] was based upon resources available and retained only within the continenta United States."

One of the Powells' attorneys Abraham J. Wirin of Los Angeles, spen eight weeks in Communist China i 1958 gathering information for the tria When he returned, he told newsmer "Yes, I have seen evidence of germ war fare on the part of this country, but I ar not expressing an opinion that this ev dence is true. The truth of the evidenc will have to be decided by the jury." H added, however, that he had "secure evidence which, if introduced at th trial and accepted by the jury, woul constitute a complete defense to th charges . . . sufficient evidence to su port a verdict of the jury acquitting th defendants."

Years later, Wirin told a differer story. "I am far from convinced that th United States engaged in biological war fare," he said in an interview in Jul 1967. "The results of the germ warfar were, in my opinion, so minimal that w wouldn't have taken the chance." Wiri added that he had gone to England t see Dr. Joseph Needham, one of th members of the 1952 Internation Commission, but "he wouldn't subjec himself to a cross-examination. H wouldn't come to the United States" t testify.

Germ warfare charges against th United States were raised again, thi time by Cuba, in June, 1964. A com munique handed newsmen in Prim Minister Fidel Castro's presidential pa lance announced that the regime wa investigating a "possible U.S.-instigate germ warfare attack" the previou week. "There would be extraordinaril grave and unpredictable consequence if this is true," the communique addec Castro did not appear. U.S. officia scoffed at the Cuban charges, describ ing them as "absurd and preposterous. Some expressed the view that the Cu ban outburst was part of a campaign t counter anti-Castro propaganda emanat of the field manual. On another page, the manual explicitly states the current U.S. doctrine: "The United States is not a party to any treaty, now in force, that prohibits or restricts the use in warfare of toxic or nontoxic gases, or smoke or incendiary materials or of bacteriological warfare."

In a private letter dated May, 1964, Paul R. Ignatius, then Under Secretary of the Army, noted that "these weapons [CBW], like the nuclear-armed missiles and bombs we have poised around the world, are responsive only to the orders of the President. . . . " Ignatius added that the weapons would be used only to counter an attack from an aggressor, but such a stipulation does not appear in Army Field Manual (FM) 101-40, *Armed Forces Doctrine for Chemical and Biological Weapons Employment and Defense.*

This manual, which is accepted doctrine for all four services, says that "the decision for U.S. forces to use chemical and biological weapons rests with the President of the United States." Commanders receive directives relating to the employment of CB munitions through normal chain of command channels. The pattern and objectives for the use of CB agents, says FM 101-40, will depend upon such variables as U.S. foreign policy, requirements of the military situation, the participation of allies, nature of the enemy, and related factors. Once the decision has been made to conduct CB operations, authority to use CB weapons is normally delegated to the commander who has responsibility for the attack area.

The manual nowhere indicates that U.S. policy is to strike only in retaliation. At one point, it allows that "under specific favorable conditions, chemical agents may be employed with a minimum of advance planning to allow the attack of fleeting targets." Otherwise, all attacks must be carefully coordinated between the White House, the Secretary of Defense, the Joint Chiefs of Staff and military field units.

One month after the *New York Times* printed the 1964 dispatch about the growing CBW effort, U.S. soldiers in South Vietnam directed the first use of tear and nausea gas against Viet Cong guerrillas, attacks that violated no U.S. military policies. Yet when news of such warfare became known with worldwide repercussions in March, 1965, Secretary of State Dean Rusk ignored that aspect and argued instead that the 1925 Geneva convention had not ruled out non-lethal riot control gases similar to those used in South Vietnam. Rusk and other top-level Johnson Administration officials recalled the 1943 Roosevelt statement and indicated that this was still U.S. policy.

As late as May, 1967, then Deputy Secretary of Defense Cyrus R. Vance told a Senate disarmament subcommittee that the United States was actively developing chemical and biological weapons, but would never initiate their use. After noting the Roosevelt statement, he told the Senators:

"We think we must have a retaliatory capability and a defensive capability [in CBW], and those are the ends to which we are devoting both our research and development and our procurement. It is clearly our policy not to initiate the use of lethal chemicals or lethal biologicals.[7]

[7]The recent emphasis on the war-time speech is ironic, because there is must evidence that Roosevelt's statement carried little weight with his contemporaries. Shortly before he died in 1966, Fleet Admiral Chester W. Nimitz was asked by a reporter what his toughest World War II decision had been. The old admiral answered: "There were a lot of tough ones. I think when the War Department suggested the use of poison gas during the invasion of Iwo Jima, that was a trying decision. I decided the United States should not be the first to violate the Geneva Convention. It cost many fine Marines." And in the second volume of his autobiography, published in 1964, David E. Lilienthal, former Chairman of the Atomic Energy Commission (AEC), reported that the United

Public statements and assurances aside, the Defense and State departments have bitterly fought any attempts to convert the Roosevelt statement into official U.S. policy.

In 1959, Representative Robert W. Kastenmeier, Wisconsin Democrat, alarmed by a wave of Army-planted magazine articles extolling the humanity of CBW, began a long, unsuccessful battle to have the White House and Pentagon endorse the Roosevelt no-first-use statement. On September 3, he introduced a concurrent resolution calling for the Administration to reaffirm "the longstanding policy of the United States that in the event of war the United States shall under no circumstances resort to the use of poisonous or obnoxious gases unless they are first used by our enemies." In a floor speech that day, the Congressman warned that the Chemical Corps was seeking a change in the Roosevelt policy. "We should not accede to the judgment of the certain military officers who want the right to use chemical and biological weapons as pre-emptive attack weapons. ... I feel that now is the time that we must re-evaluate, before it is too late, where we are going in the field of CBW."

Kastenmeier also argued that in the absence of a no-first-use declaration by the United States, the Soviet Union and other nations that had ratified the Geneva Accords were scoring propaganda victories by quoting some of the statements made by chemical corps generals and others and claiming that American CBW preparations were intended for aggressive warfare use. He argued that the nation would lose the propaganda battle "unless we are willing to express publicly a moral national policy on this issue."

(At about this time, according to former staff aids of the Wisconsin Congressman, he received a call from two high-ranking Chemical Corps officers. They elaborately explained to Kastenmeier that the use of CBW was humane; the Congressman responded by offering to tone down his resolution. Instead of calling for a restatement of policy concerning no-first-use of any CBW agents, Kastenmeier said he would be willing to make the restriction apply to only the first use of lethal chemical agents such as nerve gas. The Chemical Corps officers refused the compromise and explained that the decision of whether or not to use lethal or non-lethal agents would have to be made on the battlefield.)

President Eisenhower talked about the U.S. policies on CBW at his news conference on January 13, 1960—the first time since 1943 that the question of CBW had been publicly discussed by a President. Asked if the Defense Department was considering a change away from the no-first-use policies, Eisenhower answered: "I will say this: no such official suggestion has been made to me and so far as my own instinct is concerned it is to not start such a thing as that first."

But by this time, as we have seen, military policy toward first use of CBW had

States was prepared to use poison gas against the Japanese late in World War II. He said a recommendation to that effect had been prepared by the late General George C. Marshall while he was U.S. Army Chief of Staff. Lilienthal further quoted Marshall as explaining that "the reason it [gas] was not used was chiefly the strong opposition of Churchill and the British. They were afraid that this would be the signal for the Germans to use gas against England." Thus, the fear of retaliation, and not respect for the 1925 Geneva accord or the 1943 Roosevelt pledge, presumably kept the United States from being the first nation to use gas warfare in World War II. Admiral William D. Leahy, in his 1950 autobiography, *I Was There*, also reveals that the United States was prepared to use bacteriological warfare against Japanese rice crops if necessary. He particularly recalled one such discussion with President Roosevelt as they were sailing for Honolulu for the MacArthur-Nimitz conferences in July of 1944. As Leahy put it: "Personally, I recoiled from the idea and said to Roosevelt, "Mr. President, this [using germs and poisons] would violate every Christian ethic I have ever heard of and all of the known rules of law. It would be an attack on the noncombatant population of the enemy. The reaction can be foretold—if we use it, the enemy will use it."

already changed, according to the Army manuals and General Creasy's 1958 Congressional testimony. Although the facts are cloaked in secrecy, the National Security Council apparently reviewed the Pentagon's new CBW directives sometime in the late 1950's and ruled that such weapons could not be used except in retaliation. Since the new ruling did not limit CBW retaliation specifically to initial attacks with CBW agents, this move had the effect of bringing the CBW policies in line with the then-existing directives toward the initial use of nuclear weapons, which permitted strikes of "massive retaliation" even if the enemy attacked first with only conventional weapons.

Another indication that the CBW arsenal had been incorporated into the Pentagon's overall weapons systems, albeit with Security Council restrictions, came on March 29, 1960, when the Defense Department urged that Kastenmeier's resolution not be approved by the House Foreign Affairs Committee, before which it was pending. The Defense Department's statement said, in part:

"Similar declarations might apply with equal pertinency across the entire weapons spectrum, and no reason is perceived why biological and chemical weapons should be singled out for this special declaration. Whether the use of any major type of weapon should be initiated is a matter to be decided at the highest levels of Government in the light of the Nation's longstanding policies and principles, its international obligations, and the emergent situations it will confront. Effective controls on biological and chemical weapons, as in the case of other weapons, may have to await international agreements with necessary safeguards."

After arguing that there was "evidence" that the Communist nations were actively pursuing CBW programs, the Pentagon statement said the proposed resolution appeared "to introduce uncertainty into the necessary planning of the Department of Defense in preparing to meet possible hostile actions of all kinds."

The State Department also attacked Kastenmeier's resolution. In a letter to the Committee on April 11, 1960, it said:

"We must recognize our responsibilities toward our own and the free world's security. These responsibilities involve, among other things, the maintenance of an adequate defense posture across the entire weapons spectrum, which will allow us to defend against acts of aggression in such a manner as the President may direct. Accordingly, the Department believes that the resolution should not be adopted.

Thus the Pentagon and State Department each refused an opportunity to renew the Roosevelt policy of no-first-strike use of CBW agents, a policy they insisted in 1965 had not been altered by the initial use of chemical warfare in South Vietnam. The U.S. position was that the Roosevelt pledge—and the 1925 Geneva Accords—did not deal with non-lethal gases, such as the tear and nausea agents being used by American forces in Vietnam. Thus no violation had occurred, in the U.S. view.

9.

The Scientists Appeal for Peace

Linus Pauling

Linus Pauling, Nobel laureate, is one of those rare scientist-humanists who occasionally appear on the historical scene. He won the Nobel Prize in proton chemistry, followed, years later, by the Nobel Peace Prize.

Professor Pauling projects his brilliant mind on the problems of human survival in his book *No More War!*

He once stated:

"Science is the search for the truth—it is not a game in which one tries to beat his opponent, to do harm to others. . . . I believe in morality, in justice, in humanitarianism. We must recognize now that the power to destroy the world by the use of nuclear weapons is a power that cannot be used—we cannot accept the idea of such monstrous immorality.

"The time has now come for morality to take its place in the conduct of world affairs; the time has now come for the nations of the world to submit to just regulation of their conduct by international law."

Many statements about nuclear weapons and war have been made by scientists since the nuclear age began.

Even before the first nuclear bomb had been tested it became clear to the scientists working on the project that the new weapons were a herald of man's mastery of an entirely new world of overwhelming forces, and that a new kind of statesmanship would be needed to steer mankind away from disaster.

A memorandum directed to President Roosevelt was written by Dr. Leo Szilard in March 1945. In this memorandum it was pointed out that a system of international control of nuclear weapons might give us a chance of living through this century without having our cities destroyed. President Roosevelt died before this memorandum reached him. On 28 May 1945, six weeks before the first bomb was tested in New Mexico, Dr. Szilard discussed this memorandum in a personal interview with Mr. James F. Byrnes, to whom it had been referred by the White House. The full text of the Szilard Memorandum has not yet been published; excerpts were printed in the December 1947 issue of the *Bulletin of the Atomic Scientists*.

A report on "Social and Political Implications of Atomic Energy," called the Franck Report, was transmitted to the Secretary of War on 11 June 1945. It had been prepared by a committee of three physicists, three chemists, and one biologist: Drs. James Franck (chairman), Donald Hughes, Leo Szilard, Thorfin Hogness, Glenn Seaborg, Eugene Rabinowitch, and J. J. Nickson. Its primary purpose was to warn against the use of the atomic bomb against Japan. It contained the statements "We believe that these considerations make the use of nuclear bombs for an early unannounced attack against Japan inadvisable. If the United States were to be the first to release this new means of indiscriminate destruction on mankind, she would sacrifice public support throughout the world, precipitate the race for armaments, and prejudice the possibility of reaching an international agreement on the future control of such weapons."

The Franck Report was published in the *Bulletin of the Atomic Scientists* on 1 May 1946.

Since 1945 there have been many appeals by scientists and by laymen for sanity in the nuclear world—so many that they cannot all be reproduced here.

On 15 July 1955 a declaration was issued calling on all nations to renounce force as a final resort of policy.

At the same time, July 1955, the Russell-Einstein appeal was issued. This appeal, which was formulated by Bertrand Russell and was signed by Einstein a few days before his death, pointed out the dangers of thermonuclear weapons. It contained the statement "There lies before us, if we choose, continual progress in happiness, knowledge and wisdom. Shall we, instead, choose death, because we cannot forget our quarrels? We appeal, as human beings, to human beings: remember your humanity and forget the rest. If you can do so, the way lies open to a new paradise; if you cannot, there lies before you the risk of universal death."

The signers of the appeal (Max Born, P.W. Bridgman, Albert Einstein, L. Infeld, F. Joliot-Curie, H. J. Muller, Linus Pauling, C. F. Powell, J. Rotblat, Bertrand Russell, and Hideki Yukawa) asked that an international congress of scientists be convened, and urged that they pass the following resolution:

"In view of the fact that in any future world war nuclear weapons will certainly be employed, and that such weapons threaten the continued existence of mankind, we urge the Governments of the world to realize, and to acknowledge publicly, that their purposes cannot be furthered by a world war, and we urge them, consequently, to find peaceful means for the settlement of all matters of dispute between them."

In response to this appeal a conference (the First Pugwash Conference) was held in Pugwash, Nova Scotia, in July 1957. A stirring and informative report was prepared, and it was signed by twenty scientists, of ten nations. The report covered the hazards arising from the use of atomic energy in peace and war, the problem of the control of nuclear weapons, and the social responsibility of scientists.

Three Soviet scientists were among those at Pugwash. On their return to Moscow they made a report to the Academy of Sciences of the U.S.S.R., and they prepared a statement supporting the Pugwash report, ending with the sentence: "We Soviet scientists express our full readiness for common effort with scientists of any other country, to discuss any proposals directed toward the prevention of atomic war and the creation of secure peace and tranquility for all mankind." This statement was signed by 198 members of the Academy of Sciences and other Soviet academies.

The Second Pugwash Conference was held in April 1958, and plans were made for a larger one to be held in September 1958.

Among other appeals by scientists, that of 13 April 1957 signed by 18 leading physicists of West Germany is noteworthy in that these scientists not only urged West Germany to renounce nuclear weapons of all kinds, but also stated that they would not take part in the production, testing, or use of nuclear weapons.

On 24 April 1957 the Declaration of Conscience by Dr. Albert Schweitzer was broadcast from Oslo.

The Scientists' Petition to the United Nations

At noon on Monday 15 January 1958

I placed in the hands of Mr. Dag Hammarskjold, Secretary-General of the United Nations, a petition from 9235 scientists, of many countries in the world.

This petition has the title "Petition to the United Nations Urging that an International Agreement to Stop the Testing of Nuclear Bombs be Made Now."

The petition consists of five paragraphs, as follows:

"We, the scientists whose names are signed below, urge that an international agreement to stop the testing of nuclear bombs be made now.

"Each nuclear bomb test spreads an added burden of radioactive elements over every part of the world. Each added amount of radiation causes damage to the health of human beings all over the world and causes damages to the pool of human germ plasm such as to lead to an increase in the number of seriously defective children that will be born in future generations.

"So long as these weapons are in the hands of only three powers an agreement for their control is feasible. If testing continues, and the possession of these weapons spreads to additional governments, the danger of outbreak of a cataclysmic nuclear war through the reckless action of some irresponsible national leader will be greatly increased.

"An international agreement to stop the testing of nuclear bombs now could serve as a first step toward a more general disarmament and the ultimate effective abolition of nuclear weapons, averting the possibility of a nuclear war that would be a catastrophe to all humanity.

"We have in common with our fellow men a deep concern for the welfare of all human beings. As scientists we have knowledge of the dangers involved and therefore a special responsibility to make those dangers known. We deem it imperative that immediate action be taken to effect an international agreement to stop the testing of all nuclear weapons."

The letter that was given to Mr. Hammarskjold together with the petition and the names of the 9235 scientists reads as follows:

"Sir:

"On behalf of 9235 scientists of many countries of the world, I submit herewith the accompanying petition urging that an international agreement to stop the testing of nuclear bombs be made now, as a first step toward a more general disarmament.

"The petition is submitted by my 9234 fellow scientists and me as individuals. No organization has been responsible for the planning or writing of the petition or for the collection of signatures.

"The petition resulted from an address on Science in the Modern World given by me in the Chapel of Washington University, St. Louis, Missouri, to the students and faculty of the University on May 15, 1957. The response to this address was so enthusiastic as to suggest that a statement be prepared to which American scientists could subscribe. I wrote the statement, which is the accompanying petition, on that day, and within a few days twenty-six other U.S. scientists had signed: Barry Commoner, Edward U. Condon, Charles D. Coryell, Leslie C. Dunn, Viktor Hamburger, Michael Heidelberger, I. H. Herskowitz, Herbert Jehle, Martin Kamen, Edwin C. Kemble, I. M. Kolthoff, Chauncey Leake, S. E. Luria, Max Mason, Carl V. Moore, Philip Morrison, Hermann J. Muller, Severo Ochoa, C. C. Price, Arthur Roberts, M. L. Sands, Verner Schomaker, Laurence H. Snyder, Oswald Veblen, M. B. Visscher, W.H. Zachariasen. Within two weeks the signatures of over 2000 U.S. scientists had been obtained and on the fourth of June 1957 I submitted the statement to the President of the United States of America.

"In July 1957 I received a voluntary

statement of adherence to the petition from all of the professors of science (forty in number) of the Free University of Brussels, and similar statements from scientists in several other countries. I later wrote to a few scientists in every country, asking that they sign the petition and obtain other signatures.

"So far the petition has been signed by 9235 scientists, of 44 countries. Among the signers are 36 Nobel Laureates, representing 12 countries. Their names are given in the first list (and repeated in the main list). Separate lists are also given of 101 members of the Academy of Sciences of the United States of America, 35 Fellows of the Royal Society of London, England, and 216 Members and Correspondents of the Academy of Sciences of the U.S.S.R.

"In some countries only a few distinguished scientists have signed the petition; an example is Sweden, with two signers, both Nobel Laureates. In others (Japan, Rumania) the effort seems to have been made by my correspondents to request nearly all of the scientists of the nation to sign. Sir C. V. Raman, of India, a Nobel Laureate in physics, wrote that in his opinion every scientist in India would sign the petition if he had the opportunity; the number of Indian signers, 535, should not be taken to mean that many scientists are opposed. It is my opinion that the petition represents the feelings of the great majority of the scientists of the world.

"We urge that an international agreement to stop the testing of nuclear weapons be made now. The details of the agreement should, of course, be such that it be effective, and that, so far as possible, it benefit all nations and all people equally, not one nation or group of nations preferentially. We urge that this agreement be made as a first step toward a more general disarmament, and we hope that later steps will be taken without delay.

"It is not my belief that this problem is one that should be settled by scientists; it is instead one of importance to every person in the world. My colleagues and I feel, however, that it is worthwhile for us to express our opinion to you, as we have done with this petition, inasmuch as it is the scientists who have some measure of understanding of the complex factors involved in the problem, such as the magnitude of the genetic and somatic effects of the released radioactive materials. This may be illustrated by the statement of one of the signers of the petition, J. H. Burn, F.R.S., Professor of Pharmacology in the University of Oxford, that we are particularly concerned about the role of uptake in the bones of children of radioactive strontium from the milk they drink.

"We are also, of course, greatly concerned about the danger of outbreak of a cataclysmic war. We believe that international problems should be solved not by war, but by the application of man's power to reason—through arbitration, negotiation, international agreements, international law—and that a just and effective international agreement to stop bomb tests would be a good first step."

Respectfully yours,

LINUS PAULING

It is mentioned in the letter that there were 36 Nobel Laureates among the signers. The signature of one additional Nobel Laureate was added to the list later. These 37 Nobel Laureates, of thirteen countries, are all scientists, although two of them, Albert Schweitzer and Lord Boyd Orr, received their Nobel prizes in peace, and one, Bertrand Russell, received his in literature. Of the others, nine received Nobel prizes in physics, twelve in chemistry, and thirteen in physiology and medicine.

Eight of the Nobel Laureates are Americans, seven are British, eight are German, two are French, two are Chinese (both residing now in the United States, where they made the discovery that led to the award of the Nobel prize to them), two are Swedish, two Swiss, one Irish, one Japanese, one Indian, one Hungarian (now resident in the United States), one Belgian, and one Russian.

The names of the Nobel Laureates who signed the petition are given in the following list:

Nobel Laureates Among Signers

Nobel Laureates in Physics

Max Born
Germany
P. A. M. Dirac
Great Britain
W. Heisenberg
Germany
Tang-Dao Lee
(Resident in U.S.A.)
China
E. T. S. Walton
Ireland
C. F. Powell
Great Britain
C. V. Raman
India
C. N. Yang
(Resident in U.S.A.)
China
Hideki Yukawa
Japan

Nobel Laureates in Chemistry

K. Alder
Germany
A. Butenandt
Germany
Otto Hahn
Germany
Leopold Ruzicka
Switzerland
N. N. Semenov
U.S.S.R
R. L. M. Synge
Great Britain
Frederic Joliot-Curie
France
Richard Kuhn
Germany
Linus Pauling
U.S.A.
A. W. K. Tiselius
Sweden
Harold Urey
U.S.A.
Adolf Windaus
Germany

Nobel Laureate in Literature

Bertrand Russell
Great Britain

Nobel Laureates in Peace

Lord Boyd Orr
Great Britain
Albert Schweitzer
France

Nobel Laureates in Physiology and Medicine

Jules Bordet
Belgium
Henry Dale
Great Britain
W. P. Murphy
U.S.A.
A. Szent-Györgyi
(Resident in U.S.A.)
Hungary
Gerhard Domagk
Germany
Joseph Erlanger
U.S.A.
Hans Krebs
Great Britain
Otto Loewi
U.S.A.
Hermann Muller
U.S.A.
Max Theiler
U.S.A.
Hugo Theorell
Sweden
T. Reichstein
Switzerland
G. H. Whipple
U.S.A.

Copies of the petition and of a list of names of some of the signers were delivered to all of the 82 national missions to the United Nations on the same day on which the petition was submitted.

I asked the Permanent Mission of India to the United Nations to circulate the petition, for they had repeatedly spoken out against nuclear tests since 1954. However, while they were waiting for their government's permission, the U.S.S.R. delegation, without consulting me and probably without knowing of India's interest in the petition, made a formal request of the Secretary-General that they be allowed to circulate the petition, and they did so on 28 February 1958.

In the meantime, many additional signatures had arrived in Pasadena, and a second list of names of scientists who had signed the petition brought the total number of scientists signing the petition to 11,021, representing 49 countries.

The petition was the result of the efforts of individual scientists. No organization was responsible for circulating the petition and gathering signatures. The whole job was done by a rather few people, and the entire expense, which was not very great, was borne by a few individuals, all scientists, except some volunteers who contributed some secretarial labor and one Pasadena resident who sent me a check for $100 to help defray the expenses.

I think that the ease with which the many thousands of signatures were obtained reflects the strong desire of scientists all over the world to contribute something to the solution of the urgent and all-important world problem that has been posed by the testing and stockpiling of the immense stores of nuclear weapons.

The Origin of the Appeal by American Scientists

For many years I have been concerned about the contradiction that exists between the ethical principles of behavior that apply to individual human beings, and are in general conformed to, and the immorality of the actions of nations and of national leaders, who are willing, instead of settling their differences in a moral and peaceful way, to sacrifice the lives of millions of human beings.

When I first heard that atomic bombs had been exploded over Hiroshima and Nagasaki in 1945 I was shocked, as were many other scientists, by the terible powers of destruction that had been made available to man by the progress of science. Like many other scientists, I began to talk to groups of people—to luncheon clubs, to labor unions, to groups of students, to peace meetings—about the catastrophe that would come to the world if there were to be a war in which atomic bombs were used. Many times since 1945 I have talked to such groups about the overwhelming importance to the world of preserving peace.

On 26 March 1957 I spoke at the Santa Barbara Conference on World Affairs about the great problem that the world now faces, the problem of averting a nuclear war fought with hydrogen bombs and superbombs. It seemed to me that the citizens of Santa Barbara and the other participants in this Conference were more keenly aware than before of the possibility of a world catastrophe, but that there was considerable confusion and uncertainty in their minds as to the opinions of scientists, because of the apparent conflict between statements that had been published by scientists of the Atomic Energy Commission, and some others personally connected with the war effort, and those made by other scientists.

In his powerful radio address "A Declaration of Conscience," broadcast on 23 April 1957 from Oslo to most of the countries of the world, Dr. Albert Schweitzer pointed out that fallout radioactivity from the tests of great nuclear weapons causes damage to the health of human beings now living and to our descendants. He called for an expression of informed public opinion in all nations that would lead the statesmen to reach an agreement to stop the bomb tests, and concluded by saying that "The end of further experiments with atom bombs would be like the early sunrays of hope which suffering humanity is longing for."

To this appeal by the great humanitarian, an answer was at once sent by AEC Commissioner Dr. Willard F. Libby. The answer contains statements that seem to be worded in such a way as to suggest to the reader that fallout does not cause damage: " . . . exposures from fallout are very much smaller than those which would be required to produce observable effects in the population." It contains no estimate of the number of people in the world who would be affected by fallout radioactivity, such as a scientist with more facts about the matter at his disposal than any other scientist in the world would be expected to make, but only reassurances and vague mention of extremely small risks, undetectable effects, hazards very much less than those that persons take as a normal part of their lives. This answer to Dr. Schweitzer and other statements made by Dr. Libby seemed to me to indicate that the AEC was not willing to trust the public with the facts of the true hazard.

Then on 15 May 1957 I spoke on the campus of Washington University in St. Louis. I had been invited by the faculty committee in charge to give an address in the Graham Memorial Chapel of the University on the occasion of the Eliot Honors Day Assembly, held in honor of the students who had been elected to Phi Beta Kappa, Sigma Xi, Order of the Coif, and other honor societies. The subject of my address was Science and the Modern World. I talked about how many wonderful aspects of the world had been found through the investigations of scientists. I talked about the discoveries of physicists—about electrons, the nuclei of atoms, protons, neutrons, neutrinos and antineutrinos; and also about the structure of molecules, about the way in which some human beings manufacture abnormal molecules of hemoglobin, which causes them to have a disease, a kind of hereditary hemolytic anemia. They manufacture these abnormal molecules because they have inherited bad genes from both father and mother. I mentioned that another bad gene, the gene for phenylketonuria, causes one percent of the cases of mental deficiency that are found in our mental hospitals. I mentioned that the bomb tests now being made are increasing the number of bad genes, and are probably also causing people to die of leukemia and other diseases. Then I said that Dr. Albert Schweitzer has said that "A humanitarian is a man who believes that no human being should be sacrificed to a project." I continued in the following way: "I am a humanitarian. I believe that no human being should be sacrificed to a project; and in particular I believe that no human being should be sacrificed to the project of perfecting nuclear weapons that could kill hundreds of millions of human beings, could devastate this beautiful world in which we live." I ended my address with a quotation from a letter written in 1780 by Benjamin Franklin to the scientist Joseph Priestly: "The rapid progress true Science now makes occasions my regretting sometimes that I was born so soon. It is impossible to imagine the height to which may be carried, in a thousand years, the power of man over matter. O that *Moral* science were in as fair a way of improvement, that men would cease to be wolves to one another, and that human beings would at length learn what they now improperly call *humanity.*"

The applause that came at the end of this address and the many questions as to whether there were some action that they could take that were asked me by some of the thousand students and members of the faculty of Washington University who were present made me decide to go ahead with the idea of preparing an appeal that could be signed by American scientists. This idea had arisen in a discussion that I had had the day before when I was telling Dr. Barry Commoner, Professor of Botany in Washington University and chairman of the committee that had invited me to give the Chapel address, what I planned to say in my address.

The Appeal by American Scientists

That afternoon I wrote the appeal, with the help of some of the Washington University scientists. Its wording was exactly the same as that of the petition to the United Nations, given earlier in this chapter, except that it began "We, the American scientists whose names are signed below . . . " Its heading was "An Appeal by American Scientists to the Governments and People of the World."

That evening some copies of the appeal were mimeographed and some letters were typed, which I sent to a few other scientists, asking if they would join me in making the appeal. Within a week I had received answers from 26 of them; their names are listed in the letter to Mr. Hammarskjold that has been reproduced above.

On 22 May a few hundred copies of the appeal were printed and mailed to people in various universities and scientific laboratories in the United States. By ten days later, 2 June, the signatures of 2000 American scientists had reached me at my home in Pasadena. Many of the copies of the appeal that were returned have ten or twenty or thirty signatures attached.

It is significant that most of the leading geneticists of the country signed the appeal. It is, of course, the geneticists who understand best the damage that radiation can do to the human race by its action on the genes, and who are most concerned about this aspect of the bomb tests and of nuclear war.

On 3 June 1957 a statement about the Appeal by American Scientists was made to the press. A copy of the appeal, with a statement about it and the names of many of the signers, was also sent to the Special Subcommittee on Radiation of the Joint Committee on Atomic Energy of the Congress of the United States, which was holding hearings at that time. A letter and a copy of the Appeal were also sent to President Eisenhower.

Public mention of the Appeal by American Scientists was made by President Eisenhower in his press conference of 6 June 1957. When he was asked about his comment by Mr. James Reston of *The New York Times*, the President replied, "I said that there does seem to be some organization behind it [the Appeal by American Scientists]. I didn't say a wicked organization."

However, the President was mistaken in his surmise. The Appeal by American Scientists did not have any organization behind it; it was purely the result of private initiative and private enterprise on the part of individual scientists.

I believe that the Appeal by American Scientists in June 1957 did much good for the United States, in helping to correct some mistaken ideas about the United States and about American scientists that seem to have gained credence in foreign countries. From 9 June until 1 September 1957 my wife and I were in Europe, where we attended scientific congresses in several countries. Several scientists in these countries mentioned to us that the Appeal by American Scientists showed that American scientists are as concerned as those of other countries about the great problem of nuclear war and the related problem of the nuclear bomb tests.

The Origin of the Petition to the United Nations

The Appeal by American Scientists grew into the petition by scientists to the United Nations in a gradual and unexpected way.

On 8 June 1957, a few days after public announcement had been made about the Appeal by American Scientists, I was asked by a reporter in New York if scientists of other countries might join the appeal. I said "I should like to see the signatures of thousands of Russian scientists, of British and French scientists, of scientists of all countries of the world, to this appeal," and this statement was published in some newspapers.

A month later I received the following communication from the professors of science in the Free University of Brussels:

"We, the undersigned members of the staff of the Free University of Brussels, associate ourselves with the Appeal by Professor Pauling and other American scientists urging that an international agreement be made now to stop the testing of all nuclear weapons."

It bore the signatures of 40 Belgian scientists.

Shortly thereafter several similar communications reached me from individual scientists and small groups of scientists in other countries, and on my return to Pasadena in September I decided that it would be worthwhile to ask scientists in other nations of the world to associate themselves with the Appeal by American Scientists. I employed a personal secretary for this purpose, and wrote about 500 letters to scientists in foreign countries. Many of these letters were to scientists whom I had met in international scientific meetings or on visits to foreign countries, or who had visited the California Institute of Technology in Pasadena. Others were addressed to scientists whose names were known to me because of their authorship of scientific papers. Other names were selected from reference books, especially the book *The World of Learning,* which lists all of the professors in the universities and learned institutions in the world.

These 500 letters were well recived; they resulted in my getting about 7500 signatures of scientists in countries other than the United States.

This is a yield of 15 signatures per letter.

On 12 February 1958 the columnist Fulton Lewis, Jr., in his column published in many papers, asked: "Which organizations or individuals helped Dr. Pauling in his worldwide operation? Also, such a petition costs money; lots of it. Experts tell me that the expense would average $10 per signature. It is certainly not amiss for a congressional committee to inquire who raised the necessary $100,000."

Who are Mr. Lewis's experts? They seem not to understand this issue—they greatly overestimated the cost of getting the signatures of scientists to an appeal urging an international agreement to stop bomb tests. It was about three cents per signature, instead of ten dollars.

However, I have been told by people with experience in gathering signatures by mail that a yield of one signature in ten letters is not unusual in such a campaign. The yield that I obtained, 15 signatures per letter, was accordingly 150 times the ordinary yield. I think that this result can be interpreted only as meaning that the scientists of the world are indeed greatly concerned about the problem of nuclear testing and nuclear war, and that this great concern is responsible for their enthusiastic cooperation.

The cost of gathering the 7500 signatures of scientists outside the United States amounted to about $250.00, which I expended for stationery, postage, and secretarial help. I am happy to be able to make this small contribution to the solution of world problems. I am, of course, pleased that the signatures were obtained at the bargain rate of about three cents per signature.

My wife and I have expended altogether about $600 on the appeal and petition, for gathering signatures, preparing material for press conferences and lists for distribution to the National Missions to the U.N., etc. This sum covered most of the cost of the whole job.

In December I was invited by the American Nobel Anniversary Committee to be present at their annual dinner, to be held in the Waldorf-Astoria Hotel in New York on the evening of 11 January 1958. I was asked also to take part in the program that evening, together with several other Nobel Laureates: Lord Boyd Orr of Great Britain, Mr. Lester Pearson of Canada, Miss Pearl Buck, and Dr. Albert Szent-Györgyi. Dr. Clarence Pickett, who as Executive Secretary of the American Friends Service Committee had received the Nobel Prize for Peace that was given to that organization, was also a participant in the program. I accepted the invitation.

I had been wondering what to do with the 9000 signatures of scientists, from all over the world, that had reached me by that time. It occurred to me that the thing to do with the appeal was to present it as a petition to the United Nations.

On arrival in New York I telephoned the United Nations, asked Mr. Hammarskjold's secretary for an appointment with Mr. Hammarskjold, and explained my purpose. I was given the appointment for noon on 13 January 1958, and at that time my wife and I visited Mr. Hammarskjold and presented the petition to him.

The Signers of the Petition

Among the signers of the petition are many of the leading scientists of the world.

A list of the 37 Nobel Laureates who signed the petition has already been given. The names of a few of them are repeated in the following listings.

The National Academy of Sciences of the United States of America is a distinguished organization, founded by an Act of Congress in 1864. It has, according to its charter, the obligation to advise the government of the United States about scientific matters, when requested by the government to do so.

The National Academy of Sciences has about 500 members. Of these a total of 105, one fifth, are among the signers.

Ninety-five Fellows of the Royal Society of London, which lists among its members all of the leading scientists of the British Commonwealth, signed the petition. Among this group are Sir Henry Dale, a former president of the Royal Society and of the Royal Institution of Great Britain, and Sir Charles Darwin, former director of the National Physical Laboratory.

Among the signers of the petition were 216 leading scientists of the Soviet Union. This group includes Professor N. Semenov, Nobel Laureate in chemistry in 1956; also Professor A. Nesmeyanov, the distinguished chemist who is President of the Academy of Sciences of the U.S.S.R., Professor A. Topchiev, the chemist who is Secretary of the Academy of Sciences, Professor P. Kapitza, well known physicist who is a Foreign Associate of the National Academy of Sciences of the United States and a Fellow of the Royal Society of London, and many other scientists with world-wide reputations.

Number of signers of the Petition, by Countries

Argentina	56	Japan	1,161
Australia	8	Jordan	11
Australia	1	Lebanon	1
Belgian Congo	6	Mexico	23
Belgium	43	Netherlands	29
Brazil	152	New Zealand	119
Bulgaria	392	Norway	112
Burma	50	Panama	2
Canada	24	Peru	1
Ceylon	27	Poland	86
China	4	Portugal	1
Colombia	34	Rumania	2,749
Czechoslovakia	284	South Africa	41
Denmark	53	Spain	7
Ecuador	13	Sweden	2
Egypt	236	Switzerland	13
France	463	Thailand	20
Germany	151	Turkey	5
Ghana	1	Uruguay	9
Great Britain	701	U.S.A.	2,875
Greece	1	U.S.S.R	216
India	535	Venezuela	19
Ireland	9	West Pakistan	1
Israel	57	Yugoslavia	38
Italy	179		
		Total	11,021

The numbers of the signers in each of the countries represented, as of May 1958, are given in the following list. There is an interesting variability in the response from the foreign countries, which probably is accidental in its origin—dependent upon the nature of the response of the usually three or four scientists in each country to whom my letters were addressed.

10.

The Myth of the Peaceful Atom

by Richard Curtis and Elizabeth Hogan

Coauthors Elizabeth Hogan and Richard Curtis joined their respective talents to conduct a thorough study of the ramifications of the commercial uses of atomic energy. Their investigation led to consultations with nuclear scientists, engineers, and biologists.

The Myth of the Peaceful Atom is beautifully written, thought provoking, sometimes frightening, and sprinkled with touches of controversy.

"What is past is past, and the damage we may already have done to future generations cannot be rescinded, but we cannot shirk the compelling responsibility to determine if the course we are following is one we should be following."

So said Senator Thruston B. Morton of Kentucky on February 29, 1968, upon introducing into Congress a resolution calling for comprehensive review of federal participation in the atomic energy power program. Admitting he had been remiss in informing himself on this "grave danger," Morton said he had now looked more deeply into nuclear power safety and was "dismayed at some of the things I have found—warnings and facts from highly qualified people who firmly believe that we have moved too fast and without proper safeguards into an atomic power age."

Senator Morton's resolution on nuclear power was by no means the only one before Congress in 1968. Indeed, more than two dozen legislators urged investigation and reevaluation of this program. This fact may come as a surprise to much of the public, for the belief is widespread that the nuclear reactors being built to generate electricity for our cities are safe, reliable, and pollution-free. But a rapidly growing number of physicists, biologists, engineers, public health officials, and even staff members of the Atomic Energy Commission itself—the government bureau responsible for regulation of this force—have been expressing serious misgivings about the planned proliferation of nuclear power plants. In fact, some have indicated that nuclear power, which Supreme Court Justices William O. Douglas and Hugo L. Black described as "the most deadly, the most dangerous process that man has ever conceived," represents the gravest pollution threat yet to our environment.

As of June, 1968, 15 commercial nuclear power plants were operating or operable within the United States, producing about one per cent of our current electrical output. The government, however, has been promoting a plan by which 25 per cent of our electric power will be generated by the atom by 1980, and half by the year 2000. To meet this goal, 87 more plants are under construction or on the drawing boards. Although atomic power and reactor technology are still imperfect sciences, saturated with hazards and unknowns, these reactors are going up in close proximity to heavy population concentrations. Most of them will be of a size never previously attempted by scientists and engineers. They are in effect gigantic nuclear experiments.

As most readers will recall, atomic reactors are designed to use the tremendous heat generated by splitting atoms. They are fueled with a concentrated form of uranium stored in thin-walled tubes bound together to form subassemblies. These are placed in the reactor's core, separated by control rods that absorb neutrons and thus help regulate chain reactions of splitting atoms. When the rods are withdrawn, the chain reactions intensify, producing enormous quantities of heat. Coolant circulated through the fuel elements in the reactor core carries the heat away to heat-exchange systems, where water is brought to a boil. The resultant steam is employed to turn electricity-generating turbines.

Stated in this condensed fashion, the process sounds innocuous enough. Unfortunately, however, heat is not the only form of energy produced by atomic fission. Another is radioactivity. During the course of operation, the fuel assemblies and other components in the reactor's core become intensely radioactive. Some of the fission by-products have been described as a million to a billion times more toxic than any known industrial chemical. Some 200 radioactive isotopes are produced as by-products of reactor operation and the amount of just one of them, strontium-90, accumulated in a reactor of even modest (100–200 megawatt) size, after it has been operative for six months, is equal to what would be produced by the explosion of a bomb 190 times more powerful than the one dropped on Hiroshima.

Huge concentrations of radioactive material are also to be found in nuclear fuel-reprocessing plants. Because the intense radioactivity in a reactor core eventually interferes with the fuel's efficiency, the spent fuel assemblies must be removed from time to time and replaced by new, uncontaminated ones. The old ones are transported to reprocessing plants where the contaminants are separated from the salvageable fuel as well as from plutonium, a valuable by-product. Since no satisfactory means have been found for neutralizing or for safely releasing into the environment the radioactive liquid containing the contaminants, it must be stored until it is no longer dangerous. Thus, reprocessing plants and storage areas are immense repositories of "hot" and "dirty" material. Furthermore, routes between nuclear power plants and the reprocessing facility carry traffic bearing high quantities of such material.

Even from this glimpse it will be apparent that public and environmental safety depend on the flawless containment of radioactivity every step of the way. For, owing to the incredible potency of fission products, even the slightest leakage is harmful and a massive release would be catastrophic. The fundamental question, then, is how heavily can we rely on human wisdom, care, and engineering to hold this peril under absolute control?

Abundant evidence points to the conclusion that we cannot rely on it at all.

The hazards of peaceful atomic power fall into two broad categories: the threat of violent, massive releases of radioactivity or that of slow, but deadly, seepage of harmful products into the environment.

Nuclear physicists assure us that reactors cannot explode like atomic bombs because the complex apparatus for detonating an atomic warhead is absent. This fact, however, is of little consolation when it is realized that only a *conventional* explosion, which ruptures the reactor mechanism and its containment structure, could produce havoc on a scale eclipsing any industrial accident on record or any single act of war, including the atomic destruction of Hiroshima or Nagasaki.

There are numerous ways in which

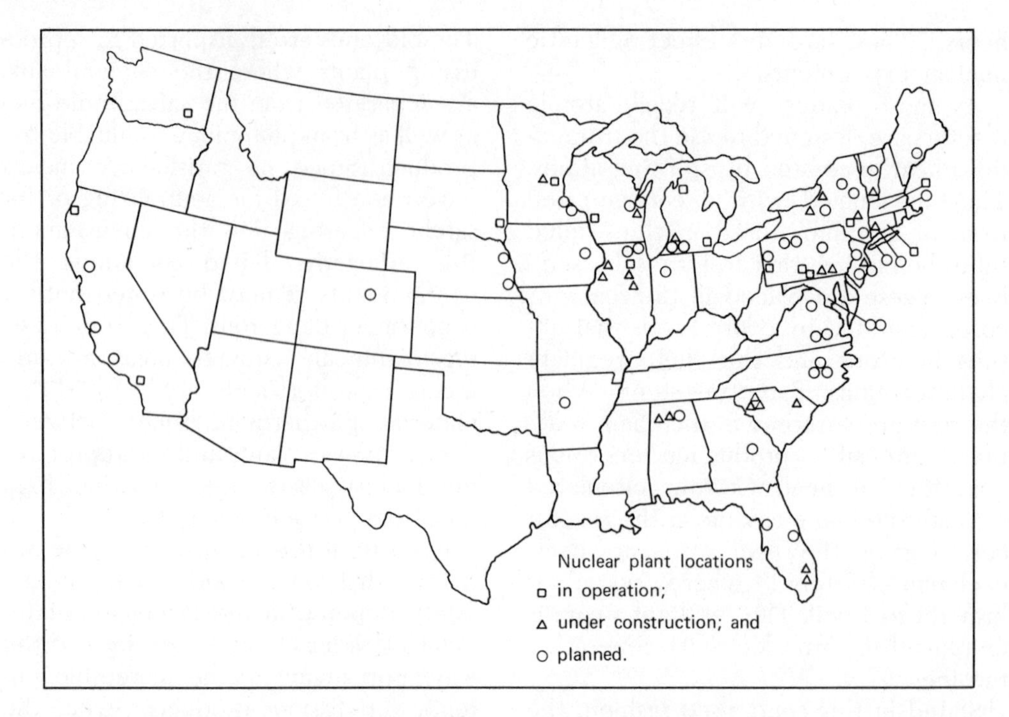

such an explosion can take place in a reactor. For example, liquid sodium, which is used in some reactors as a coolant, is a devilishly tricky element that under certain circumstances burns violently on contact with air. Accidental exposure of sodium could initiate a chain of reactions: rupturing fuel assemblies, damaging components and shielding and destroying primary and secondary emergency safeguards. If coolant is lost, as it could be in some types of reactors, fuel could melt and recongeal, forming "puddles" that could explode upon reaching a critical size. If these explosions are forceful enough, and safeguards fail, some of the fission products could be released outside the plant and into the environment in the form of a gas or a cloud of fine radioactive particles. Under not uncommon atmospheric conditions such as an "inversion," in which a layer of warm air keeps a cooler layer from rising, a blanket of radioactivity could spread insidiously over the countryside. Another possibility is that fission products could be carried out of the reactor and into a city's watershed, for all reactors are being built on lakes, rivers, or other bodies of water for cooling purposes.

What would be the toll of such a calamity?

In 1957 the Atomic Energy Commission issued a study (designated Wash.—740), largely prepared by the Brookhaven National Laboratory, that attempted to assess the probabilities of such "incidents" and the potential consequences. Some of its findings were stupefying: From the explosion of a 100–200 megawatt reactor, as many as 3,400 people could be killed, 43,000 injured, and as much as 7 billion dollars of property damage done. People could be killed at distances up to 15 miles and injured up to 45. Land contamination could extend for far greater distances: agricultural quarantines might prevail over an area of 150,000 square miles, more than the combined areas of Pennsylvania, New York, and New Jersey.

The awful significance of these figures is difficult to comprehend. By way of comparison, we might look at one of the worst industrial accidents of modern

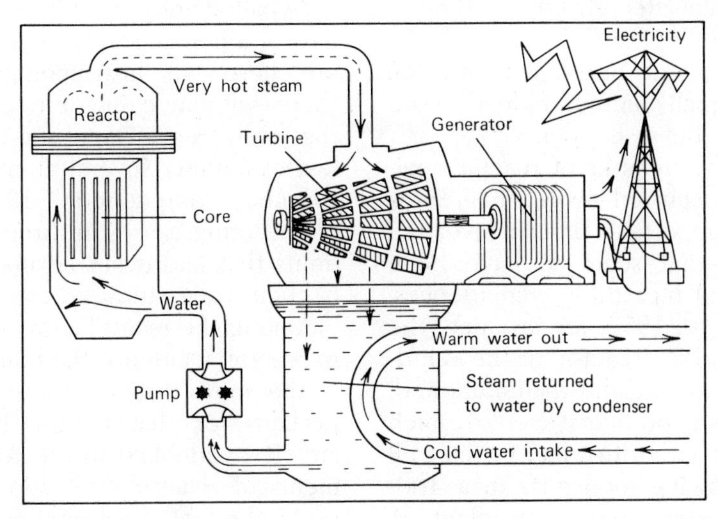

Boiling water reactor.

times: the Texas City disaster of 1947 when a ship loaded with ammonium nitrate fertilizer exploded, virtually leveling the city, killing 561 people, and causing an estimated $67 million worth of damage. Appalling as this catastrophe was, however, it does not begin to approach the potential havoc that would be wreaked by a nuclear explosion occurring in one of the plants now being constructed close to several American cities.

The scientists and engineers who produced the Brookhaven Report optimistically ventured to give high odds against such an occurrence, asserting that the structures, systems, and safeguards of atomic plants were so engineered as to render it practically incredible. At the same time, though, the report was replete with such statements as:

"The cumulative effect of radiation on physical and chemical properties of materials, after long periods of time, is largely unknown."

"Much remains to be learned about the characteristics and behavior of nuclear systems."

"It is important to recognize that the magnitudes of many of the crucial factors in this study are not quantitatively established, either by theoretical and experimental data or adequate experience."

Even if the report had been founded on more substantial understanding of natural and technical processes, any of the grounds on which the Brookhaven team based its conclusions are shaky at best.

For one thing, all of us are familiar with technological disasters that have occurred against fantastically high odds: the sinking of the "unsinkable" *Titanic*, or the November 9, 1965, "blackout" of the northeastern United States, for example. The latter happening illustrates how an "incredible" event can occur in the electric utility field, most experts agreeing that the chain of circumstances that brought it about was so improbable that the odds against it defy calculation.

Congressional testimony given in 1967 by Dr. David Okrent, a former chairman of the AEC's Advisory Committee on Reactor Safeguards, demonstrated that fate is not always a respecter of enormously adverse odds. "We do have on record cases where, for example, an applicant, appearing before an atomic safety and licensing board, stated that a mathematical impossibility had occurred; namely, one tornado took out five separate power lines to a reactor. If one calculated strictly on the basis of probability and multiplied the probability for one line five times, you get a very small number indeed," said Dr. Okrent, "but it happened."

A disturbing number of reactor accidents have occurred—with sheer luck playing an important part in averting catastrophe—that seem to have been the product of incredible coincidences. On October 10, 1957, for instance, the Number One Pile (reactor) at the Windscale Works in England malfunctioned, spewing fission products over so much territory that authorities had to seize all milk and growing foodstuffs in a 400-square-mile area around the plant. A british report on the incident stated that *all* of the reactor's containment features had failed. And, closer to home, a meltdown of fuel in the Fermi reactor in Lagoona Beach, Michigan, in October, 1966, came within an ace of turning into a nuclear "runaway." An explosive release of radioactive materials was averted, but the failures of Fermi's safeguards made the event, in the words of Sheldon Novick in *Scientist and Citizen,* "a bit worse than the 'maximum credible accident.' "

The atomic industry has attempted to design components and safeguards so that failure of one vital system in a plant will not affect another, resulting in a "house of cards" collapse. However, two highly regarded authorities, Theos J. Thompson and J. G. Beckerley, in a book on reactor safety advise us not to place too much faith in claims of independent safeguards: "A structure as complex as a reactor and involving as many phenomena is likely to have relatively few completely independent components." Many manufacturers and utility operators have resisted the idea of producing "redundant safeguards" on the grounds of excessive cost.

Investigations of reactor breakdowns usually disclose a number of small, seemingly unrelated failures, which snowballed into one big one. A design flaw or a human error, a component failure here, an instrumentation failure there—all may coincide to contribute to the total event. Thompson and Beckerley, examining several atomic plant accidents, pinpointed 13 different contributing causes in three of the accidents that had occurred up to the time of their 1964 study.

Among the many factors contributing to reactor accidents, the human element is the most difficult to quantify. And perhaps for that reason, it has been largely overlooked in the AEC's assessments of reactor safety. Yet, a private researcher of nuclear accidents, Dr. Donald Oken, M.D., Associate Director of the Psychosomatic and Psychiatric Institute of Michael Reese Hospital in Chicago reported: "A review of reports of past criticality and reactor incidents and discussions held with some of the health personnel in charge reveal a number of striking peculiarities in the behavior of many of those involved—in which they almost literally asked for trouble."

AEC annuals are full of reports of human negligence: 3,844 pounds of uranium hexafluoride lost owing to an error in opening a cylinder; a $220,000 fire in a reactor because of accidental tripping of valves by electricians during previous maintenance work; numerous vehicular accidents involving transport of nuclear materials. None of these accidents led to disaster, but who will warrant that, with the projected proliferation of power plants and satellite industries in the coming decade, a moment's misjudgment will not trigger a nightmare? Perhaps worse, the likelihood of sabotage has scarcely been weighed, despite a number of incidents and threats.

It should be apparent that if men are to build safe, successful reactors, the whole level of industrial workmanship, engineering, inspection, and quality control must be raised well above prevailing levels. The more sophisticated the technology, the more precise the correspondence between the subtlest gradations of care or negligence and that technology's success or failure. When meters, grams, and seconds are no longer good enough, and specifications call for millimeters, milligrams, and milliseconds, the demands made on men, materials, and machinery are accordingly intensified. Minute lapses that might be tolerable in a conventional industrial procedure will wreck the more exacting one. And when the technology is not only exacting but hazardous in the extreme, then a trivial oversight, a minor defect, a moment's inattention may spell doom.

While there is little doubt that American technology is the most refined on earth, there is ample reason to believe that it has more than met its match in the seemingly insurmountable problems posed by the peaceful atom. Societies of professional engineers, and others concerned with establishing technical and safety criteria for the nuclear industry, have described between 2,800 and 5,000 technical standards that are necessary for a typical reactor power plant in such areas as materials, testing, design, electrical gear, instrumentation, plant equipment, and processes. Yet, due to the rapidity with which the nuclear industry has developed, as of March, 1967, only 100 of these had been passed on and approved for use.

It is not surprising, then, to learn that serious technical difficulties are turning up in reactor after reactor. At the Big Rock Point Nuclear Plant, a relatively small reactor near Charlevoix, Michigan, control rods were found sticking in position, studs failing or cracked, screws jostled out of place and into key mechanisms, a valve malfunctioning for more than a dozen reasons, foreign material lodging in critical moving parts, and welds cracked on every one of sixteen screws holding two components in place. A reactor at Humboldt Bay in California manifested cracks in the tubes containing fuel: in order to keep costs down, stainless steel had been used instead of a more reliable alloy. The Oyster Creek plant in New Jersey showed cracks in 123 of 137 fuel tubes, and welding defects at every point where tubes and control-rod housings were joined around the reactor's vessel. Reactors in Wisconsin, Minnesota, Connecticut, Puerto Rico, New York, and elsewhere have experienced innumerable operating difficulties, and some, such as the $55 million Hallam plant in Nebraska have been forced to shut down for good, owing to plant malfunction.

Chilling parallels can be drawn between failures in nuclear utility technology and in the nuclear submarine program. In October, 1962, Vice Admiral Hyman G. Rickover, Director of AEC's Division of Naval Reactors, took the atomic industry to task in a speech in New York City:

"It is not well enough understood that conventional components of advanced systems must necessarily meet higher standards. Yet it should be obvious that failures that would be trivial if they occurred in a conventional application will have serious consequences in a nuclear plant because here radioactivity is involved. . . ."

Rickover went on to cite defective welds, forging materials substituted without authorization, violations of official specifications, poor inspection techniques, small and seemingly "unimportant" parts left out of components, faulty brazing of wires, and more. "I assure you," he declared, "I am not exaggerating the situation; in fact, I have understated it. For every case I have given, I could cite a dozen more."

The following April, the U.S. atomic submarine *Thresher*, while undergoing a deep test dive some 200 miles off the Cape Cod coast, went down with 112 naval personnel and 17 civilians and never came up again. Subsequent investigation revealed that the sub suffered from many of the same ailments described in Rickover's speech. "It is extremely unfortunate," said Senator John O. Pastore, chairman of the joint congressional committee that held hearings on the disaster, "that this tragedy had to occur to bring a number of unsatisfactory conditions into the open." We must now ask if the same will one day be said about a power plant near one of our large cities.

If a major reactor catastrophe did occur there is good reason to believe that the consequences would be far worse than even the dismaying toll suggested by the 1957 Brookhaven Report, for a number of developments since then have made the threat considerably more formidable.

The Brookhaven Report's accident statistics, for instance, pertained to a reactor of between 100 and 200 megawatts. But while the 15 reactors currently operating in the United States average about 186 megawatts, the 87 plants going up or planned for the next decade are many times that size. Thirty-one under construction average about 726 megawatts; 42 in the planning stage average 832; 14 more, planned but without reactors ordered, will average 904. Some, such as those slated for Illinois, California, Alabama, and New York anticipate capacities of more than 1,000 megawatts. Con Edison has just announced it intends to build four units of 1,000 megawatts each on Long Island Sound near New Rochelle in teeming Westchester County—four nuclear reactors, each with a capacity five to ten times that of the reactor described in the Brookhaven Report.

These facilities will accordingly contain more uranium fuel, and because it is costly to replace spent fuel assemblies (this delicate and dangerous process can take six weeks or longer), the new reactors are designed to operate without fuel replacement far beyond the six months posited in the Brookhaven Report. As a result, the buildup of toxic fission products in tomorrow's reactors will be far greater than at present, and an accident occurring close to the end of the "fuel cycle" in such a plant could release fantastic amounts of radioactive material.

Most serious of all, perhaps, is that tomorrow's reactors are now slated for location in close proximity to population concentrations. While the Brookhaven Report had its hypothetical reactor situated about 30 miles from a major city, many of tomorrow's atomic plants will be much closer. Although the AEC has drafted "guidelines" for siting reactors, the Commission has failed to make utilities adhere to them. In 1967, Clifford K. Beck, AEC's Deputy Director of Regulation, admitted to the Joint Committee on Atomic Energy that nuclear plants in Connecticut, California, New York, and other locations, "have been approved with lower distances than our general guides would have indicated when they were approved."

Also, we must remember that while a reactor may not be near the legal boundaries of a metropolis, it may lie close to a population center. Thus, while Con Edison's Indian Point plant is 24 miles from New York City (two more plants are now being built there), it is within 10 miles of an estimated population of 155,510. It need only be recalled that the Brookhaven Report foresaw people being killed by a major radioactive release at distances up to 15 miles to realize the significance of these figures.

In a recent study of nuclear plant siting made by W. K. Davis and J. E. Robb of San Francisco's Bechtel Corporation, the locations of 42 nuclear power plants (some proposed, some now operable) were examined with respect to population centers inhabited by 25,000 residents or more. Their findings are unnerving: only *two* plants in operation or planned are more than 30 miles from a population center. Of the rest, 14 are between 20 and 27 miles away, 15 between 10 and 16 miles, and 11 between 1 and 9 miles.

Is it necessary to build atomic plants so big and so close? The answer has to do with economics. The larger a facility is, the lower the unit cost of construction and operation and the cheaper the electricity. The longer the fuel cycle, the fewer the expensive shutdowns while spent fuel assemblies are replaced. The closer the plant is to the consumer, the lower the cost of rights of way, power lines, and other transmission equipment.

On a few occasions an aroused public has successfully opposed the situation of plants near population centers. When the Pacific Gas and Electric Company persisted in trying to build a reactor squarely over earthquake faults in an area of known seismic activity—the site was Bodega Head, north of San Francisco—a courageous conservation group forced the company to back down. It has been suggested, though, that the group might not have won had not the Alaskan earthquake of 1964, occurring while the fight was going on, underscored the recklessness of the utility's scheme.

Announcement by Con Edison at the end of 1962 of its proposal to build a large nuclear plant in Ravenswood, Queens, close to the center of New York City brought a storm of frightened and angry protest. Although the utility's chairman noted, "We are confident that a nuclear plant can be built in Long Island City, or in Times Square for that matter, without hazard to our own employees working in the plant or to the community," David E. Lilienthal, the former head of the AEC, had a contrary opinion, declaring he "would not dream of living in Queens if a huge nuclear plant were located there." Outraged citizens and a number of noted scientists prevailed.

For the most part, however, the battle has been a losing one. Con Edison, for example, after its defeat in the Ravenswood fight, has just announced an interest in building a reactor on Welfare Island, literally a stone's throw from midtown Manhattan. Also, New York's Governor Nelson Rockefeller has gone on record advocating an $8 billion electric power expansion program based extensively on nuclear energy. The state legislature approved of the program, and in 1968, voted to bolster the plan with state subsidies.

Some of the deepest concern about the size and location of atomic plants has been expressed by members of the AEC themselves. "The actual experience with reactors in general is still quite limited," said Harold Price, AEC's Director of Regulation, in 1967 congressional hearings, "and with large reactors of the type now being considered, it is nonexistent. Therefore, because there would be a large number of people close by and because of lack of experience, it is . . . a matter of judgment and prudence at present to locate reactors where the protection of distance will be present."

Beck's statement is mild compared to that made in the same hearings by Nunzio J. Palladino, Chairman of the AEC's Advisory Committee on Reactor Safeguards for 1967, and Dr. David Okrent, former Chairman for 1966: "the ACRS believes that placing large nuclear reactors close to population centers will require considerable further improvements in safety, and that *none of the large power reactors now under construction is considered suitable for location in metropolitan areas* [our italics]."

The threat of a nuclear plant catastrophe constitutes only half of the double jeopardy in which atomic power has placed us. For even if no such calamity occurs, the gradual exhaustion of what one scientist terms our environmental "radiation budget," due to unavoidable releases of radioactivity during normal operation of nuclear facilities, poses an equal and possibly more insidious threat to all living things on earth.

Most of the fission products created in a reactor are trapped. Contaminated solids, liquids, and gasses are isolated, allowed to decay for a short period of time, then concentrated and shipped in drums to storage areas. These are called "high level wastes." But technology for retaining all radioactive contaminants, is either unperfected or costly, and much material of low-level radioactivity is routinely released into the air or water at the reactor site. These releases are undertaken in such a way, we are told, as to insure dispersion or dilution sufficient to prevent any predictable human exposure above harmful levels. Thus, when atomic power advocates are asked about the dangers of contaminating the environment, they imply that the relatively small amounts of radioactive materials released under "planned" conditions are harmless.

This view is a myth.

In the first place, many waste radionuclides take an extraordinarily long time to decay. The half-life (the time it takes for half of an element's atoms to disintegrate through fission) of strontium-90, for instance, is more than 27 years. Thus, even though certain long-lived isotopes are widely dispersed in air or diluted in water, their radioactivity does not cease. It remains, and over a period of time accumulates. It is therefore not pertinent to talk about the safety of any single release of "hot" effluents into the environment. At issue, rather, is their duration and cumulative radioactivity.

Further, many radioactive elements taken into the body tend to build up in specific tissues and organs to which those isotopes are attracted, increasing by many times the exposure dosage in those local areas of the body. Iodine-131, for instance, seeks the thyroid gland; strontium-90 collects in the bones; cesium-137 accumulates in muscle. Many isotopes have long half-lives, some measurable in decades.

Two more factors controvert the view that carefully monitored releases of low-level radioactivity into the environment are not pernicious. First, there is apparently no radiation threshold below which harm is impossible. Any dose, however small, will take its toll of cell material, and that damage is irreversible. Second, it may take decades for organic damage, or generations for genetic damage, to manifest itself. In 1955, for example, two British doctors reported a case of skin cancer—ultimately fatal—that had taken forty-nine years to develop following fluoroscopic irradiation of a patient.

Still another problem has received inadequate attention. Man is by no means the only creature in whom radioactive isotopes concentrate. The dietary needs of all plant and animal life dictate intake of specific elements. These concentrate even in the lowest and most basic forms of life. They are then passed up food chains, from grass to cattle to milk to man, for example. As they progress up these chains, the concentrations often increase, sometimes by hundreds of thousands of times. And if these elements are radioactive. . . .

Take zinc-65, produced in a reactor when atomic particles interact with zinc in certain components. Scrutiny of the wildlife in a pond receiving runoff from the Savannah River Plant near Aiken, South Carolina, disclosed that while the water in that pond contained only infinitesimal traces of radioactive zinc-65, the algae that lived on the water had concentrated the isotope by nearly 6,-

000 times. The bones of bluegills, an omnivorous fish that feeds both on algae and on algae-eating fish, showed concentrations more than 8,200 times higher than the amount found in the water. Study of the Columbia River, on which the Hanford, Washington, reactor is located revealed that while the radioactivity of the water was relatively insignificant: 1. the radioactivity of the river plankton was 2,000 times greater; 2. the radioactivity of the fish and ducks feeding on the plankton was 15,000 and 40,000 times greater, respectively; 3. the radioactivity of young swallows fed on insects caught by their parents near the river was 500,000 times greater; and 4. the radioactivity of the egg yolks of water birds was more than a million times greater.

Here then are clear illustrations of the ways in which almost undetectable traces of radioactivity in air, water, or soil may be progressively concentrated, so that by the time it ends up on man's plate or in his glass it is a tidy package of poison.

That nuclear facilities are producing dangerous buildups of radio-isotopes in our environment can be amply documented. University of Nevada investigators, seeking a cause for concentrations of iodine-131 in cattle thyroids in wide areas of the western United States, concluded that "the principal known source of I-131 that could contribute to this level is exhaust gases from nuclear reactors and associated fuel-processing plants."

In his keynote address to the Health Physics Society Symposium at Atlanta, Georgia, early in 1968, AEC Commissioner Wilfred E. Johnson admitted that the release into the atmosphere of tritium and noble gases such as krypton-85 would present a potential problem in the future, and that, as yet, scientists had not devised a way of solving it. Krypton-85, although inert, has a 10-year half-life and tends to dissolve in fatty tissue, meaning fairly even distribution throughout the human body. Krypton-85 is particularly difficult to filter out of reactor discharges, and the accumulation of this element alone may exhaust as much as two-thirds of the "average" human's "radiation budget" for the coming century, based on the standards established by the National Committee on Radiation Protection and Measurement.

That "low-level" waste is a grossly deceptive term is obvious. In his book *Living with the Atom*, author Ritchie Calder in 1962 described an "audit" of environmental radiation that he and his colleagues, meeting at a symposium in Chicago, drew up to assess then current and future amounts of radioactivity released into atmosphere and water. Speculations covered the period 1955–65, and because atomic power plants were few and small during that time, the figures are more significant in relation to the future. Tallying "planned releases" of radiation from such sources as commercial and test reactors, nuclear ships, uranium mills, plutonium factories, and fuel-reprocessing plants, Calder's group came to a most disquieting conclusion: "By the time we had added up all the curies which might predictably be released, by all those peaceful uses, into the environment, it came to about 13 million curies per annum." A "curie" is a standard unit of radioactivity whose lethality can be appreciated from the fact that one trillionth of one curie of radioactive gas per cubic meter of air in a uranium mine is ten times higher than the official maximum permissible dose.

Calder's figures did not include fallout due to bomb testing and similar experiments, nor did they take into account possible reactor or nuclear transportation accidents. Above all, they did not include possible escape of stored high-level radioactive wastes, the implications of which were awesome to contemplate:

"What kept nagging us was the question of waste disposal and of the remaining radioactivity which must not get loose. We were told that the dangerous waste, which is kept in storage, amounted to 10,000 million curies. If you wanted to play 'the numbers game' as an irresponsible exercise, you could divide this by the population of the world and find that it is over 3 curies for every individual."

Exactly what does Calder mean by "the question of waste disposal"?

It has been estimated that a ton of spent fuel in reprocessing will produce from forty to several hundred gallons of waste. This substance is a violently lethal mixture of short-and long-lived isotopes. It would take five cubic miles of water to dilute the waste from just *one* ton of fuel to a safe concentration. Or, if we permitted it to decay naturally until it reached the safe level—and the word "safe" is used advisedly—just one of the isotopes, strontium-90, would still be damaging to life 1,000 years from now, when it will have only one seventeen-billionth of its current potency.

There is no known way to reduce the toxicity of these isotopes; they must decay naturally, meaning *virtually perpetual containment.* Unfortunately, mankind has exhibited little skill in perpetual creations, and procedures for handling radioactive wastes leave everything to be desired. Formerly dumped in the ocean, the most common practice today is to store the concentrates in large steel tanks shielded by earth and concrete. This method has been employed for some twenty years, and about 80 million gallons of waste are now in storage in about 200 tanks. This "liquor" generates so much heat it boils by itself for years. Most of the inventory in these caldrons is waste from weapons production, but within thirty years, the accumulation from commercial nuclear power will soar if we embark upon the expansion program now being promoted by the AEC. Dr. Donald R. Chadwick, chief of the Division of Health of the U.S. Public Health Service, estimated in 1963 that the accumulated volume of waste material would come to two billion gallons by 1995.

It is not just the volume that fills one with sickening apprehension but the techniques of disposing of this material. David Lilienthal put his finger on the crux of the matter when he stated: "These huge quantities of radioactive wastes must somehow be removed from the reactors, must—without mishap—be put into containers that will never rupture; then these vast quantities of poisonous stuff must be moved either to a burial ground or to reprocessing and concentration plants, handled again, and disposed of, by burial or otherwise, with a risk of human error at every step." Nor can it be stressed strongly enough that we are not discussing a brief danger period of days, months, or years. We are talking of periods "longer," in the words of AEC Commissioner Wilfred E. Johnson, "than the history of most governments that the world has seen."

Yet already there are many instances of the failure of storage facilities. An article in an AEC publication has cited nine cases of tank failure out of 183 tanks located in Washington, South Carolina, and Idaho. And a passage in the AEC's authorizing legislation for 1968 called for funding of $2,500,000 for the replacment of failed and failing tanks in Richland, Washington. "There is no assurance," concluded the passage, "that the need for new waste storage tanks can be forestalled." If this is the case after twenty years of storage experience, it is beyond belief that this burden will be borne without some storage failures for centuries in the future. Remember too, that these waste-holding "tank farms" are vulnerable to natural catastrophes such as earthquakes, and to man-made ones such as sabotage.

Efforts are of course being made toward effective handling of the waste problems, but many technical barriers must still be overcome. It is unlikely they will all be overcome by the end of the century, when waste tanks will boil with 6 billion curies of strontium-90, 5.3 billion curies of cesium-137, 6.07 billion curies of prometheum-147, 10.1 billion curies of cerium-144, and millions of curies of other isotopes. The amount of strontium-90 alone is 30 times more than would be released by the nuclear war envisioned in a 1959 congressional hearing.

The burden that radioactive wastes place on future generations is cruel and may prove intolerable. Physicist Joel A. Snow stated it well when he wrote in *Scientist and Citizen:* "Over periods of hundreds of years it is impossible to ensure that society will remain responsive to the problems created by the legacy of nuclear waste which we have left behind."

"Legacy" is indeed a gracious way of describing the reality of this situation, for at the very least we are saddling our children and their descendants with perpetual custodianship of our atomic refuse, and at worst may be dooming them to the same agonizing afflictions and deaths suffered by those who survived Hiroshima. Radiation has been positively linked to cancer, leukemia, brain damage, infant mortality, cataracts, sterility, genetic defects and mutations, and general shortening of life.

The implications for the survival of mankind can be glimpsed by considering just one of these effects, the genetic. In a 1960 article, James F. Crow, Professor of Genetics at the University of Wisconsin School of Medicine and president of the Genetics Society of America, stated that for every roentgen of slow radiation—the kind we can expect to receive in increasing doses from peacetime nuclear activity—about five mutations will result per 100 million genes exposed, meaning that "after a number of generations of exposure to one roentgen per generation, about one in 8,000 . . . in each generation would have severe genetic defects attributable to the radiation."

The Atomic Energy Commission is aware of the many objections that have been raised to the atomic power program: why does it continue to encourage it? Unfortunately, the Commission must perform two conflicting roles. On the one hand, it is responsible for regulating the atomic power industry. But on the other, it has been charged by Congress to promote the use of nuclear energy by the utility industry. Because of its involvement in the highest priorities of national security, enormous power and legislative advantages have been vested in the AEC, enabling it to fulfill its role as promoter with almost unhampered success—while its effectiveness as regulator has gradually atrophied. The Commission consistently denies claims that atomic power is heading for troubled waters, optimistically reassuring critics that these plants are safe, clean neighbors.

The fact that there is no foundation for this optimism is emphasized by the insurance situation on atomic facilities. Despite the AEC's own assertion that as much as $7 billion in property damage could result from an atomic power plant catastrophe, the insurance industry, working through two pools, will put up no more than $74 million, or about one per cent, to indemnify equipment manufacturers and utility operators against damage suits from the public. The federal government will ad up to $486 million more, but this still leaves more than $6 billion in property damages to be picked up by victims of a Brookhaven-sized accident. And no insurance company—not even Lloyds of London—will issue property insurance to individuals against radiation damage. If there is so little risk in atomic power plants, why is insurance so inadequate?

The knowledge that man must henceforth live in constant dread of a major nuclear plant accident is disturbing enough. But we must recognize that even if such calamities are averted, the slow saturation of our environment with radioactive wastes will nevertheless be raising the odds that you or your heirs will fall victim to one of a multitude of afflictions. There is no "threshold" exposure below which we can feel safe.

We have little time to reflect on our alternatives, for the moment must soon come when no reversal will be possible. Dr. L. P. Hatch of Brookhaven National Laboratory vividly made this point when he told the Joint Committee on Atomic Energy:

"If we were to go on for 50 years in the atomic power industry, and find that we had reached an impasse, that we had been doing the wrong thing with the wastes and we would like to reconsider the disposal methods, it would be entirely too late, because the problem would exist and nothing could be done to change that fact for the next, say, 600 or a thousand years."

To which might be added a sobering thought stated by Dr. David Price of the U.S. Public Health Service:

"We all live under the haunting fear that something may corrupt the environment to the point where man joins the dinosaurs as an obsolete form of life. And what makes these thoughts all the more disturbing is the knowledge that our fate could perhaps be sealed twenty or more years before the development of symptoms."

What must be done to avert the perils of the peaceful atom? A number of plans have been put forward for stricter regulation of activities in the nuclear utility field, such as limiting the size of reactors or their proximity to population concentrations or building more safeguards. As sensible as these proposals appear on the surface, they fail to recognize a number of important realities: first, that such arrangements would probably be opposed by utility operators and the government due to their prohibitively high costs. Since our government seems to be committed to making atomic power plants competitive with conventionally fueled plants, and because businesses are in business for profit, it is hardly likely they would buy these answers. Second, the technical problems involved in containment of radioactivity have not been successfully overcome, and there is little likelihood they will be resolved in time to prevent immense and irrevocable harm to our environment. Third, the nature of business enterprise is unfortunately such that *perfect* policing of the atomic power industry is unachievable. As we have seen in the cases of other forms of pollution, the public spirit of men seeking profit from industrial processes does not always rise as high as the welfare of society requires. It is unwise to hope that stricter regulation would do the job.

What, then, is the answer? The only course may be to turn boldly away from atomic energy as a major source of electricity production, abandoning it as this nation has abandoned other costly but unsuccessful technological enterprises.

There is no doubt that, with this nation's demand for electricity doubling every decade, new power sources are urgently needed. Nor is there doubt that our conventional fuel reserves—coal, oil, and natural gas—are rapidly being consumed. Sufficient high-grade fossil fuel reserves exist, however, to carry us to the end of this century; and new techniques for recovering these fuels from secondary sources such as oil shale could extend the time even longer. Furthermore, advances in pollution abatement technology and revolutionary new techniques, now in development, for burning conventional fuels with high efficiency, could carry us well into the next century with the fossil fuels we have. This abundance, and potential abundance, gives us at least several decades to survey possible alternatives to atomic power, select the most promising, and develop them on an appropriate scale as alternatives to nuclear power. Solar energy, tidal power, heat from the earth's core, and even garbage and solid-waste incineration have to some degree been demonstrated as promising means of electricity generation. If we subsidized research and development of those fields as liberally as we have done atomic energy, some of them would undoubtedly prove to be what atomic energy once promised, without its deadly drawbacks.

Aside from the positive prospect of profitability in these new approaches, industry will have another powerful incentive for turning to them; namely, that atomic energy is proving to be quite the opposite of the cheap, everlasting resource envisioned at the outset of the atomic age. The prices of reactors and components and costs of construction and operation have soared in the last few years, greatly damaging nuclear power's position as a competitor with conventional fuels. If insurance premiums and other indirect subsidies are brought into line with realistic estimates of what it takes to make atomic energy both safe and economical, the atom might prove to be the most *expensive* form of energy yet devised—not the cheapest. In addition, because of our wasteful fuel policies, evidence indicates that sources of low-cost uranium will be exhausted before the turn of the century. Fuel-producing breeder reactors, in which the nuclear establishment has invested such high hopes for the creation of vast, new fuel supplies, have proven a distinct technological disappointment. Even if the problems plaguing this effort were overcome in the next ten or twenty years, it may still be too late to recoup the losses of nuclear fuel reserves brought about by prodigious mismanagement.

The proposal to abandon or severely curtail the use of atomic energy is clearly a difficult one to imagine. We have only to realize, however, that by pursuing our current civilian nuclear power program, we are jeopardizing every other industry in the country; in that light, this proposal becomes the only practical alternative. In short, the entire national community stands to benefit from the abandonment of a policy which seems to be leading us toward both environmental and economic disaster.

Man's incomplete understanding of many technological principles and natural forces is not necessarily to his discredit. Indeed, that he has erected empires despite his limited knowledge is to his glory. But that he pits this ignorance and uncertainty, and the fragile yet lethal technology he has woven out of them, against the uncertainties of nature, science, and human behavior—this may well be to his everlasting sorrow.

11.

The Dirty Animal—Man

Joseph L. Myler

Joseph L. Myler has established himself as one of the world's leading writers and reporters in the fields of space, atomic energy, and science in general.

He covered the first A-bomb test at the Bikini atoll. Ten years later, at the same spot, he covered the first U.S. air drop of a hydrogen bomb. Mr. Myler analyzes the Atomic Energy Commission's semiannual reports for United Press International in Washington, D.C., where he is a senior editor.

"The Dirty Animal, Man," is a classic of its kind. No one has written in a more articulate and meaningful fashion on the need to breathe pure air; to drink pure water; to be free from excessive noise; and the need to do something about the disposal of wastes created by humans in this technological society.

Man is poisoning his world. He has been labelled, with strong justification, the dirty animal.

He has managed to make his rivers rotten. He has transformed green pastures into deserts. He has clogged the air with chemicals which menace health and dust which is changing the climate. He is a menace to himself and other species.

He has turned large areas of his world into junk heaps, piled high or layered deep with indestructible cans or plastic containers. Americans alone discard more than a billion tons of solid waste a year and the total is growing.

Man is beginning to face up to the problem, but only slowly and against great obstacles because of government and industrial considerations.

Unless he is willing to spend billions upon billions to undo what he has done—and perhaps even change some of his basic ways—he really may be gasping for breath in a few decades.

In the developing nations, nearly a billion people get their water from unsanitary sources and half of them get sick every year as a result. Even in the United States, half the people depend on water supplies which don't meet federal standards or are of unknown quality.

Rivers of so-called developed nations have been turned into sewers of civilization to get rid of unwanted industrial wastes. The oceans are being contaminated with agricultural poisons which drain into streams and are carried away to the seas.

By exhausting warm water from our power cooling plants into the ocean, we are threatening marine life. All along we have drained the priceless topsoil of our fields into silting rivers. We have denuded many of our forests.

Millions of workers are exposed to potentially dangerous concentrations of dust, fumes, gases and vapors. No one knows what the noise generated by modern machines and cities is doing to man's nervous system.

As J. George Harrar of the Rockefeller Foundation has said, "Man himself is the greatest threat to his environment . . . we have now successfully begun to contaminate what we have not yet destroyed."

None of this happened overnight.

Once the oceans were thought to be endless, the land infinite and the atmosphere limitless. Now man's survival is known to depend on how he husbands a relatively thin layer of soil, water and air tightly wrapped around our planet's surface.

Nature with its wind and water erosion and climatic changes has been altering the environment for millions of years. But the possibility that one species might make the world uninhabitable did not arise until 8,000 years ago when the hunter, who simply ranged the land in search of food, evolved into the farmer who plowed and uprooted it.

Next came the city and then the industrial society, which multiplied the threat many times over.

Dr. Barry Commoner, Director of the Center for the Biology of Natural Systems at Washington University, said,

"Modern technology has so strained the web of the processes in the living environment at its most vulnerable points that there is little leeway left in the system.

"Unless we begin to match our technological power with a deeper understanding of the balance of nature, we run the risk of destroying this planet as a suitable place for human habitation."

Consider what man has done to one indispensable element of our biosphere—water.

Water in a sense is the most precious stuff on this planet.

Yet we waste it, we pollute it, we threaten the existence of irreplaceable underground reserves which took nature thousands of years to establish, we destroy the beauty and the life of once sparkling streams and deep blue lakes.

What is so precious about water—that cheap fluid most of us in this country can get in any amount just by turning a tap?

Throughout human history water has been the great limiter. No civilization has ever risen without a plenitude of water. When water runs out, or becomes unusable, civilizations die.

Men have killed each other for water, whether at some isolated spring in the 19th-century American West or in ancient Mesopotamia where human beings warred for control of the Tigris and Euphrates.

Water is one of the reasons for today's bloody rivalry between the Israelis and the Arabs.

The high standard of living in the United States and other affluent nations of the modern world depends on fresh water—lots of it.

Americans use about 310 billion gallons of water a day on the average for public supplies, commerce and industry, irrigation, and rural domestic and livestock needs. On a per capita basis, this is 1,600 gallons a day.

Of the annually renewable water supplies available to the United States, about 1.2 trillion gallons a day enter the streamflow from surface and underground sources.

This amount, 1.2 trillion gallons a day, constitutes the nation's ultimate water resource—for homes, industry, irrigation, recreation.

Properly managed, it can be used and reused before release to the oceans. Only a tiny amount is "consumed" in the sense that it is converted into other forms, such as chemical products, or removed as a resource by being turned into vapor.

So the United States is water rich. With all that magnificent streamflow it can never become thirsty. Or can it?

For one thing, the figures are all in averages. The blessing of fresh water from the sky ranges from less than an inch a year in some parched regions of the southwest to more than 200 inches in the Pacific Northwest and parts of Hawaii.

For another, populations and the demand for water are rising faster than man's means for making his water resources available wherever needed for human use.

The world population is expected to double to nearly 7 billion by 2000. Says Dr. Raymond L. Nace, research hydrologist of the U.S. Geological Survey:

"The problem is not whether water supplies are running out, but whether people are outrunning the supplies. Water supplies have finite limits, but the demands of people on the supplies have no known limit."

Unless he gives up piecemeal, temporary solutions to local water problems and concerns himself with the long-term global problems, man will be in trouble. For that matter, he already is.

Take pollution. To list the specific pollutants which man dumps into his water supply would take many pages.

They range from raw sewage to chemical fertilizers and animal dung, from acids and poisons generated by industry to silts and salts drained from strip mines, city streets and farmlands, from crankcase oils and detergents to disease carrying bacteria, from herbicides and pesticides to radioactive contaminants from mines and atomic plants.

Congress has enacted laws to control water pollution and is studying new ones. But the pollutant load is steadily increasing, and some of the problems involved seem almost too difficult to be solved by legislation alone.

Listen again to Nace of the Geological Survey:

"Out of its total potentially controllable liquid assets the United States uses 95 per cent chiefly as a conveyor belt on which to send waste products out to sea.

"The major use of free-running water in industrial nations is not industry, as published statistics seem to show, but waste disposal. Our rivers are open sewers."

Others have said it even more starkly.

Dr. Glenn T. Seaborg, chairman of the Atomic Energy Commission, says all 22 river systems in the United States will be "biologically dead" by the end of this century if pollution continues at present rates.

According to Hollis R. Williams of the Agriculture Department's Soil Conservation Service, "The pollution of the living waters of the United States is one of the great shames of our time."

According to former president Lyndon B. Johnson, "The clear, fresh waters that were our national heritage have become the dumping grounds for garbage and filth. They poison our fish, they breed disease, they despoil our landscapes."

These pollutants also are killing some of our lakes. Nutrients from wastes or farm fertilizers have created "Algal Blooms" which result in depletion of oxygen in the water, destroy fish, and set the stage for ultimate transformation of a lake into a marsh and eventually a meadow.

Lake Erie may already be doomed by this cycle. Lake Michigan is in danger.

According to the recent report of the Marine Science Commission, man has created a "devil's brew of pollution" which constitutes "a growing national disgrace."

How serious is all this in the world scheme? Dr. LaMont C. Cole of Cornell University has warned that mankind seems bent upon its own extinction.

The Crisis on the Horizon

Without water there could be no life of any kind on earth. In a sense water is even more precious than oxygen, the "gas of life." For without water there would be no green plants, and green plants supply the oxygen in the air we breathe.

Scientists believe life on earth originated in the primitive seas long before there was more than a trace of oxygen in the atmosphere. Oxygen, and life's dependence on it, appeared only after the evolution of plants.

The blood of animals, including man, still is a salty solution similar to sea water. The sea still surges in the circulation systems of land as well as marine creatures. Most living things are mainly water.

The sea is at once the supplier of fresh water to the land and of oxygen to the air.

More than 70 per cent of our oxygen supply, according to Cornell's Dr. LaMont Cole, comes from microscopic green plants in the sea which, like the plants of land, consume carbon dioxide with the help of solar energy and cast off oxygen as a waste product.

With his bulldozers and concrete and asphalt city-building, road-building, urbanizing man has destroyed oxygen-producing vegetation over large continental areas.

However, enough plant life, the phytoplankton, remains in the oceans to keep the oxygen content of the atmosphere fairly stable at about 20 per cent. But man is polluting even the oceans, with what consequences he does not know.

In the solar system, at least, Earth appears to be uniquely blessed with water in great quantities. Only in the case of Mars does there appear to be any faint possibility that another planet of the sun's family has or ever has had any liquid water at all.

On Earth there is a prodigious amount of water—326 million cubic miles of it. Of this hard to conceive quantity, 317 million cubic miles are in the seas which cover 71 per cent of the globe.

Most of the rest consists of "frozen assets" of fresh water locked up in glaciers and the polar icecaps.

Man, of course, is primarily concerned with available fresh water, the stuff he can drink or moisten his yards and crops with, or use in cooking, washing, and industry, or as a medium for harboring trout and other fish which it is fun to catch.

For recreation men do, of course, swarm to the sea beaches, and the estuaries, and release their tensions in many salt water sports—surf swimming and fishing, scuba diving, sailing, and lolling on the sand in the sun. They transport most of their goods in world commerce upon the salty oceans.

But the sea's great gifts to man is fresh water. The sun annually distills (evaporates) 80,000 cubic miles of fresh water from the oceans and 15,000 cubic miles from the land.

At all times about 95,000 cubic miles of water are moving between earth and sky. What goes up subsequently comes down. This, crudely put, is the hydrological, or water, cycle.

This water, as rain, snow, hail, or sleet, comes down all over the world. Most of it falls back into the oceans. But a lot of it falls on the land. The United States gets about 30 inches a year, or 4.3 trillion gallons a day. Roughly 70 per cent of this is sent back up into the air as vapor. This includes the water used by plants.

It seems silly to talk of polluting the ocean. But it is happening. DDT has been found in marine creatures everywhere. And if the plant of the ocean is jeopardized, so is the oxygen supply on which all life depends.

The Torrey Canyon oil spill of 1967 and more recent ones, including the calamity off Santa Barbara, Calif., were disasters. Animal and bird life in the spoiled areas may never be the same. Perhaps, just perhaps, these calamities were strictly local.

In any event, they might have been worse, given man's capacity for unintentional destruction. Suppose the Torrey Canyon had been loaded not with oil but with herbicides.

Cole asks the question: Would photosynthesis, the process by which plants produce oxygen, have been wiped out in the North Sea?

A few such accidents could leave man

gasping not in a matter of generations, Cole suggests, but in a matter of years.

An alarmist notion? Possibly. But those who have looked hardest at what man has done and is doing to his environment have come to expect the worst.

Some authorities hold that for the United States, at least, there is no water crisis. Says the National Academy of Sciences, "There is no nationwide shortage and no imminent danger of one."

It goes on to say, however, that "There are serious regional shortages of usable water, many of which are becoming critical because of short-sighted planning or pollution of fresh-water supplies."

Nace, of the Geological Survey, agrees that "No human crisis centering around water exists today." "But," he adds, "one is visible on the horizon."

He continues: "We have surely been living in a dream world of water abundance at prices cheaper than we pay for common dirt."

Former Secretary of Agriculture Orville L. Freeman fears that "Future generations will judge most harshly a race of men that had all the technical knowledge, all the resources they needed to provide clean water, air, and land, but lacked the will to do so."

And, says Freeman, "We are facing an environmental crisis. It affects every one of the basic elements of the biosphere—air, earth, and water, and every one of us."

Nace recently pleaded for preservation of a resource which he said is "perhaps the most valuable" humanity possesses. This is what the hydrologists call ground water. It has been stored by nature over the millennia in subterranean "aquifers" consisting of porous rock, gravel, sand, and sediments.

According to Nace, 97 per cent of all fresh liquid water on the continents is contained in aquifers which hold "many times more water than can be stored in all the surface reservoirs that will ever be built" by man. They are "buried treasure."

In arid regions they constitute the chief source of water. This nation, Nace said, need never run out of fresh water if it cooperates with nature to maintain its aquifers.

But "If these resources receive the same reckless treatment that surface waters have, we will destroy the usefulness of our only real national water resource."

Ground water supplies are menaced in many ways. They can be killed by over pumping which results in subsidence and compaction of subsurface materials to the point where they become impervious and hence useless for water storage.

They also can be made unfit for use by pollution. Encroachment of salt water into pumped coastal aquifers is one source of pollution. Septic tanks do their bit. Another source is the growing practice of underground disposal of industrial wastes.

"To sweep our wastes underground now," Nace says, "may create an unsolvable problem for the future."

One of man's new and weird pollutants is simply warm water. Most of the water taken by industry and cities from streams is used for cooling and then poured back.

According to scientists, many forms of aquatic animals and plant life are threatened by the great tonnage of heated water from power plants, nuclear or coal-fired, which is being spewed into rivers, lakes and coastal waters.

If the problem is one for the present, it is even more so for the future.

A study reported by the Geological Survey showed that in 60 of the undeveloped countries of Africa, Asia, and Latin America 90 per cent of the population depends on water supplies "that are inadequate or unsafe."

The shortage in all countries, according to Nace, is not of water but of waterworks to make the available water usable.

The United States, with tough regional water problems of its own, is trying to help less fortunate nations with theirs. In 1967 it created the Office of Water for Peace in the State Department. This agency is concerned with a host of projects ranging in scope from provision of drinkable water on a local scale to large river-basin development programs.

As part of the water for peace endeavor the United States is spending about $400 million a year in many countries to build waterworks designed to supply both household and industrial needs.

This is something that needed to be done. No nation can mature without an abundance of water. But does anybody imagine this or anything else projected will satisfy the needs of the seven billion human beings who will populate the earth by 2000 A.D. if current forecasts come true?

Nace is appalled by these population predictions in view of the inability of men "to control either nature or themselves."

"Imagine what the pollution load on water supplies could be with that many people around! Especially if the 'advanced' countries succeed in teaching the retarded ones all of their technologically ingenious ways for adding new and weird pollutants to the environment."

Coming to Terms with Nature

About 60 per cent of the earth's land area is dry or downright arid, incapable of supporting agriculture. It now accommodates only about 5 per cent of the world's people. But if the population doubles in the next few decades as predicted, these now largely empty spaces must be transformed into "living spaces."

The deserts must be made to bloom again.

Can this be done? Has mankind learned anything from the mistakes of the past, some of which actually created deserts where none existed?

In the United States we have stripped vast acreages of trees and sod. We have built great cities and greater suburbs. The result has been to speed the runoff of water and to increase the frequency of flash flooding.

Running water is power, and we use it to manufacture electricity and create huge reservoirs that are a blessing to farmers and a joy to fishermen and others seeking solace in and on the water. Not the least thing to be said of these manmade lakes is that they are beautiful.

But we also have created silting and erosion problems, and have destroyed much of the thin layer of topsoil which is the sole source of our agricultural wealth.

Not all floods, of course, are to be blamed on man. But in many regions, notably southern California, they have been aggravated by what man has done to the land.

Every year about 75,000 Americans are driven from their homes by floods. Damage is about $1 billion. As the Environmental Science Services Administration says, "Floods are also great wasters of water, and water is a priceless national resource."

To save and increase water resources and prevent or at least mitigate the effects of floods, this country has spent many billions. It can hardly be doubted that this money has been well spent.

But are we really tackling our water problems in the best way—for us and our descendants? There are those who fear we are not.

In the Far West we pump out our ground water for decades before we learn we must return unconsumed surplus water to these precious reservoirs to keep them alive. So the central valley of California is gradually sinking and killing some of its subsurface water resources.

In the southern high plains of Texas and New Mexico we are depleting underground water stores which would take nature many centuries to replenish if they were exhausted.

Dr. Barry Commoner, director of the Center for the Biology of Natural Systems at Washington University, has said mankind is suffering from an "environmental disease" created by technology.

"We are unwitting victims of environmental pollution," says Commoner, "for most of the technological affronts to the environment were made not out of greed but ignorance. .. In each case the new technology was brought into use before the ultimate hazards were known. We have been quick to reap the benefits and slow to comprehend the costs."

According to hydrologist Raymond Nace of the Geological Survey "The history of human effort to control nature is a history of continually having to combat the unwanted consequences of these efforts."

But man does have to try to solve his water problems. He has learned to recharge his aquifers with what otherwise would be water wasted as runoff.

He also is learning to conserve water resources he used to throw away.

Santee, a town near San Diego, has provided an example of what can be done to get the most out of scarce water supplies. By treating and filtering sewage, it is restoring a million gallons of water a day to drinking quality.

In Los Angeles County one purification plant reclaims 10,000 acre-feet of water a year out of a trunk sewer.

The Bethlehem Steel Co. takes the entire sewage of the city of Baltimore and, after treatment, uses it for industrial purposes. The total is more than 125 million gallons a day.

Forgetting the billions of future inhabitants of this plant for the moment, present emergencies call for immediate solutions.

Desalination of seawater is one of these. The Atomic Energy Commission has a grand thought for changing desert coastlands of the world into latter day oases.

Build huge nuclear plants capable of producing a million kilowatts of electricity and at the same time, with nuclear heat, desalting 400 million gallons of water a day.

Such a system, nicknamed a "nuplex," could produce enough power, water, and fertilizer to raise more than a billion pounds of grain a year at the site, and enough extra fertilizer to grow enough food to feed "tens of millions of people annually."

Such "agro-industrial complexes," according to the AEC, might make fit for human habitation the "one third of the world's land" which is now "virtually unoccupied—in Australia, India, Mexico, the Middle East, Peru, and the United States."

This, like Nawapa, is still only an idea. The AEC and a group of California backers had planned the world's first nuclear power-desalting plant on Bolsa Island, a manmade spit of land off the California coast near Los Angeles. It would generate 1.8 million kilowatts of power and produce 150 million gallons of fresh water daily, for an investment of $765 million.

The cost turned out to be too great, the sights were lowered, and the project is now in abeyance. No nuclear power and desalting plant has yet been built.

But desalting by other means has been undertaken. There are about 650 plants in the world capable of producing 250 million gallons of "demineralized" water a day. The biggest of them supplies Key West, Fla., with 2.6 million gallons of freshened sea water a day at prices lower than Key West used to pay for water brought by aqueduct from the Florida mainland.

Others operate in such widely separated places as Kuwait, at the head of the Persian Gulf, and Malta in the Mediterranean.

Important as these, and the far bigger plants envisaged by the AEC "nuplexes," may be on a local or regional basis, their output is a mere drop in the bucket compared to the global water supply.

If you multiplied the world's desalting plant capacity many thousands of times, according to hydrologist Nace, the "contribution to the total water supply still would be extremely small."

In other words, desalting of sea water is by no means the panacea man has been looking for. He has to look somewhere else. Even if he could convert the sea from salt to fresh water what would he do with the salt? He would be back where he started.

Where else can he look? Pending the long range solutions, when man knows more about the complexities of the hydrological cycle, what can he do?

There being no global water shortage —there's as much water in the sea as there ever was—it is necessary to do whatever can be done to ameliorate local shortages, now.

The Federal Council for Science and Technology in 1966 proposed a 10-year program for water resources research. It called for stepped up research, not overlooking the field of "far out ideas."

A decade ago a California oceanographer noted that some big icebergs from Antarctica drift far enough north to come within capture distance of interested persons above the equator.

He suggested that bergs might be towed to anchorage off Los Angeles, say, and be made to give up their locked-in fresh water for the uses of man. The notion of a crackpot?

The Federal Council warned against such unthinking dismissal of possibly good ideas. It recommended investigation "to assure that worthwhile concepts are not overlooked."

In the meantime there is nature to think about before embarking on any projects affecting truly large parts of her atmospheric, aquatic, and continental domains.

Listen again to Nace:

"Man acts for his own purposes and nature reacts according to her immutable laws. Nature, a philosopher has said, is neither friendly nor inimical. She is merely implacable . . . we had best come to terms with her. . . .

"Until we do so, our so-called conservation practices are likely to be mere tinkerings with the landscape."

Air Pollution Hangs Like an Ominous Veil

A nebulous veil hangs ominously over the earth.

Man put it there. It is polluted air, loosely called smog.

Its presence proves that man has the technical capacity to suffocate himself, and other living things on this planet.

There may be time to rend the veil, or at least to keep it from reaching the dimensions of disaster.

In many places the veil still is imperceptible to the senses. In others, its presence often is painfully apparent.

When the eyes smart and water, when breathing is a throat-burning ordeal you know—even if the sun is bright and the sky lovely—that smog is all around you.

The atmosphere is vast. For billions of years it possessed the ability to keep itself fairly clean, clean enough to sustain life on Earth.

But man has established himself as superior to nature, if only in a self-defeating sense. He can foul his environment faster than nature can clean it up.

Man has polluted his rivers and lakes and soil, and even the seas.

He has produced a crescendo of noise and clatter in his factories and offices—yes, and in his places of amusement—which not only impairs efficiency but actually threatens health.

He has strewn his highway and byways, his streams and his beaches with the ugly litter of civilization. What to do with solid waste has become gigantic puzzle with ill fitting pieces.

The atmosphere which wraps the planet in a gaseous film no thicker in proportion than an apple skin has not escaped.

In April, 1968, J. George Harrar, President of the Rockefeller Foundation, addressed the annual meeting of the National Academy of Sciences. Outside azalea bushes were flaming all over Washington, robins were chirping, all seemed right with the world. But inside the academy Harrar was saying:

"And now it is our air envelope that is endangered. The outpouring of the byproducts of modern industrialization has reached dimensions with which we are at present unable to cope.

"Incinerators, industrial and power plants, automobiles, and many other elements combine to produce a complex pollution pattern.

"As all nations increasingly industrialize, and as their cities burgeon, the possibility of eventual suffocation as the result of this pollution becomes a very real threat."

There appears to be no doubt that air pollution is a serious menace to public health. There appears to be no doubt that air pollution, unchecked, can change the world's climate, probably for the worse. If the polluted air doesn't choke us, it may trigger climatic events such as melting of the polar ice caps which would drown our greatest cities. Conversely it may hasten return of the ice age.

But there still is, say some experts, time to prevent these calamities. How much time? The Committee on Pollution of the National Academy of Sciences has reported "Air Pollution is increasing faster than our population increases."

The world population is expected to double by 2000 A.D., 31 years hence.

Another academy group, the Committee on Atmospheric Sciences, estimated that "We have, at best, but one or two generations" in which to acquire understanding of the problem and do something about it."

There are those who wonder if the effects of pollution have not already become irreversible. If they have, we live on borrowed time, and the question is how much time there is left to borrow.

Even if smog were not a threat to health and the future of the race, it is dirty and costly. Damage done by air pollution to buildings, trees, crops, livestock, materials and works of art has been estimated at $13 billion a year in the United States alone.

Even before man the atmosphere was polluted. Volcanoes have thrown dust and gases into the air from time immemorial. Other natural pollutants have been dust from wind-eroded soil, salt particles from the sea, methane from marshes, hydrogen sulfide from rotting organic matter, radioactive materials from the earth's crust, pollen, spores, and airborne bacteria.

With these natural pollutants nature could cope. The atmosphere starts at the earth's surface and rises hundreds of miles until it fades into space. It weighs a stupendous 5,700 trillion tons.

About 70 per cent of this great ocean of air—thin though it is—exists in a region about six miles high. This is the troposphere where the winds blow and the rains and snows fall and the hurricanes and tornadoes roar.

This is where the pollution, manmade or natural, tends to accumulate on the calm days. This is where nature, with wind and rain and snow does what it can to disperse or precipitate the pollutants which otherwise would clog it.

The atmosphere normally consists of about 78 per cent nitrogen, a relatively inert gas; 21 per cent oxygen, which keeps us alive; three-hundredths of 1 per cent carbon dioxide, the prime food of plants, and smaller amounts of other gases.

Now comes man with the fumes and particles released by his fires and engines and the dusts blown from the deserts he has made and the pesticides he spreads to the four winds and the solid and liquid bits of waste he dumps into the air. He threatens not only to defeat the atmosphere's self-cleansing capacity but also to change drastically its fundamental composition.

In this country we pour about 143 million tons of pollutants into the air in a single year. In thus using the atmosphere as a sort of garbage dump we unwittingly make co-conspirators of the weather and the sun.

Fog grabs man's noxious or toxic aerosols and other particulates and concentrates them in stifling pockets. It converts sulfur dioxide, a product of burning coal or oil, into sulfuric acid. Fog and sulfur oxides have figured in all the great air pollution disasters of the past.

Even the sun, the mainstay of earthly life, compounds man's felony. It plays on the hydrocarbons and nitrogen oxides from internal combustion engines and transforms them into irritating or sickening components of "photochemical smog" such as ozone, peroxyacyl nitrate (pan), and formaldehyde.

On occasion the weather creates conditions that intensify the evils already present in the atmosphere. A layer of warm air slides above the cooler surface air. This is a "temperature inversion." It keeps the dirtiest air from rising and dispersing. It traps atmospheric poisons emitted by the works of man.

Such inversions are common. Stagnant air frequently covers huge areas of the United States, particularly in fall and winter. Such times, variously called squaw winter or indian summer, can be beautiful beyond compare. They also can be dangerous for those who suffer from heart disease or the manifold respiratory maladies, asthma to emphysema, which are aggravated by smog.

Smog as a Killer

Smog is a word coined centuries ago in Britain to describe the ugly mixture of smoke and fog that sooted the countryside for miles beyond the environs of coal-burning London.

As the word is used today, smog means something much more complex than a mere mixture of smoke and fog. It is the photochemical variety that plagues Los Angeles, for example, on 200 or more days a year. Los Angeles is hemmed by a bowl of mountains which interferes with air circulation and helps to maintain stagnant inversions.

One result has been that Los Angeles leads the nation in pioneering efforts to do something about air pollution. It has at least kept the problem from getting worse for the time being.

What troubles Los Angeles and practically every other large city in the world is a chronic condition which, however unhealthy it may be, still lies somewhat this side of disaster.

It was the acute "episodes" that forced upon the world some awareness of what industrial man can do to the atmosphere and himself.

As the National Tuberculosis and Respiratory Disease Association said in a recent publication:

"It was probably the shock of the notorious air polution disasters ... that first stripped the smokestack of its glory ... and turned what was a monument to progress into a gravestone for the dead."

Some Disasters:

Meuse River Valley, Belgium. Belgium was blanketed by a thick, cold fog in the first week of December, 1930. The heavily industrialized, 15-mile-long Meuse River Valley was trapped under an inversion layer of stagnant air. Pollutants accumulated. In a few days thousands became ill; 60 died.

Donora, Pa. An inversion covered a large part of the northeastern United States in October, 1948. In the Monongahela River Valley lay Donora, a town of 14,000 crammed with industries. Nearly half of the inhabitants, 6,000, got sick. And 20 died.

London. For centuries Londoners had heated their homes with soft coal in fireplaces. London fogs were both notorious and commonplace. But in December, 1952, there occurred a fog which surpassed others of the past. A five-day spell of stagnant air killed 4,000 persons.

New York. This city's pollution disasters might have gone unnoticed if public health scientists had not studied the death records. They concluded that New York had experienced air pollution "episodes" in 1953, 1962, 1963, and 1966. In 1963, they found, 405 persons died because of poisoned air.

The chief victims in each disaster were the elderly with ailments of the heart and lungs. Weather conditions helped the polluted air to do its deadly work.

Then, as the National Tuberculosis Association points out, there have been the "industrial accidents." One foggy morning in November, 1950, in Poza Rica, Mexico, an accident at a sulfur factory resulted in a spill of hydrogen sulfide. In half an hour enough of this poison, which gives rotten eggs their loathsome stench, had been spewed into the air to sicken 320 persons and kill 22.

But there is more to air pollution than acute, sudden disaster. There is chronic daily pollution and the chronic daily damage.

The major pollutants are carbon monoxide, sulfur oxides, hydrocarbons, nitrogen oxides, particulate matter (everything from aerosols to soot), and what the official tables call "miscellaneous other."

This last category includes lead (from auto fuel), fluroides, beryllium, arsenic, asbestos, many other chemicals, and a host of pesticides, herbicides, and fungicides.

The major polluters are autos and trucks and jet airplanes. Power plants, space heating, refuse disposal (dumps and incinerators), and various industries —pulp and paper mills, iron and steel mills, oil refineries, smelters, and chemical plants.

Together, in the United States they throw 140 million tons a year of "waste" into the air, and (say the experts) the automobile is the greatest villain of all.

According to a recent Senate report the auto accounts for at least 60 per cent of total U.S. air pollution, 85 per cent of pollution in the big urban areas, 90 per cent of all carbon monoxide pollution.

In all, the auto exhausts more than 90 million tons of pollutants into the air each year—twice as much as any other fouler of the atmosphere.

If none of the fuel burned by man contained impurities (such as sulfur in coal and oil) and if all of it underwent complete combustion, the byproducts would be simply water and carbon dioxide, neither of them harmful within limits and neither listed as contaminants.

According to Dr. LaMont C. Cole of Cornell University, half of all the fuel ever burned by man has been burned in the past 50 years. Since such fuels as coal and petroleum are "a non-renewable resource," this means that man is operating an exploitive economy that "will destroy itself if continued long enough."

Moreover, his dumping of carbon dioxide into the air (six billion tons a year) threatens to change the climate and may indeed already have changed it. It also endangers the atmospheric oxygen balance which has sustained life on earth since nature invented photosynthesis, the process by which, with the help of sunlight, plants take up carbon dioxide and liberate oxygen.

Carbon dioxide is transparent to sunlight but tends to absorb heat radiated back toward space from earth. This is what meteorologists call the "greenhouse effect." It has been estimated that the planet's average temperature would be 20 degrees cooler if there were no carbon dioxide in the atmosphere.

It also has been estimated that man with his burning has increased the air's carbon dioxide content 10 to 15 per cent in the past century and will have increased it 25 per cent by the year 2000.

A worldwide warming trend was noted from 1900 to 1950. Among the effects was to move the crop line 50 to 100 miles north of the Canadian prairies. Mocking birds, once common only in the south, extended their range and their sleep-shattering song to New York.

But this trend apparently has been halted and even reversed by the other things human beings are doing to the air. Dust, smoke, and other particulates, particularly over the world's cities, are cooling the planet by reflecting sunlight away from the ground.

Doubling atmospheric carbon dioxide could increase the earth's average surface temperature nearly seven degrees. But only a 25 per cent increase in turbidity (dust, smoke, liquid particles) could lower it by the same amount.

Dr. Reid A. Bryson, University of Wisconsin climatologist, believes the veil of pollution hanging over the world already has changed its climate.

DDT dust from farm fields has been carried by winds to all corners of the earth. Soviet cities have increased smokiness over the Caucasus 19-fold since 1930. Turbidity of the air over Washington, D.C., has gone up 57 per cent in recent years.

Over Switzerland, Bryson says, it was jumped 88 per cent. In the decade between 1957 and 1967 there was a 30 per cent increase of dustiness over the Pacific Ocean. Smoky days in Chicago rose from 20 a year before 1930 to 320 in 1948.

A blue haze, probably from agricultural burning, hangs over Brazil, Southeast Asia, and Central Africa. A brown haze of dust from soils made barren by man broods over much of Africa, Arabia, India, Pakistan, and China.

The sometimes not so nebulous veil is global in extent. The jet airplane is helping to thicken it. According to Prof. W. Frank Blair of the University of Texas at Austin, the city of Dallas is visible from transport planes many miles away because, in part, of the exhausts of jet aircraft.

Jets discharge tons of particulate matter into the air every day. Their contrails, says Bryson of Wisconsin, trigger formation of high altitude cirrus clouds which tend to alter climate by reflecting sunlight back to space.

At a time when much is being said about intentional modification of the weather (which is a long way off) climatologists are worried most by inadvertent changes. This is because they can't really guess the ultimate results.

Suppose the greenhouse effect kept piling up. The ice caps would melt and the sea level would rise, perhaps as much as 400 feet. This would drown not only coastal communities but vast low lying inland areas. It also might have the effect, since the increase in water vapor would make snowfall more abundant, of starting a new ice age.

Suppose, on the other hand, that rising turbidity produced an offsetting worldwide drop in temperature. This, too, might start the glaciers crawling southward again.

Whichever way inadvertent weather modification goes, Bryson says, "It is inconceivable that the change would be beneficial."

A Miserable Mess

The evils of air pollution are more immediate than their long term influence on climate. They damage human health. They kill livestock and crops. They make the earth an unlovely habitat for living creatures.

As Dr. Harrar of the Rockefeller Foundation said, "We have known for some time that man himself is the greatest single threat to his environment." As Dr. Blair of Texas University said, "Man has made a miserable mess of the world in which he lives."

What, by fouling the air, has man done to his own health? Photochemical smog was first defined in Los Angeles in the 1940s. The city, and the state of California, have struggled mightily ever since to do something about it.

Car makers have been forced to incorporate devices which lessen dangerous emissions. Controls have been placed on factories and refuse burners. And the upshot, according to the technical weekly *Science*, is that Los Angeles is about where it was a decade ago—new sources of pollution have kept pace with the effort to curb old ones.

In a single year, according to a staff report of the Senate Commerce Committee, Los Angeles doctors advised more than 10,000 patients to move somewhere else because of the harm being done to them by air pollution.

What is this harm? Clinically it is hard to pin down because, even in the famous pollution "episodes," the victims had suffered previously from respiratory or heart maladies—the bad air just killed them sooner.

Nevertheless, according to the Senate report, air pollution "is a significant contributing factor" to the growing number of chronic diseases such as lung cancer, emphysema, bronchitis, and asthma.

City air is dirtier than country air, and statistics show that the death rate from lung cancer is 25 per cent higher in the cities than in rural regions.

A Purdue University study puts lung cancer deaths twice as high in large cities as in rural areas. It says polluted air has shortened city dwellers' lives, on a comparative basis, by five years.

In New York, New Jersey, California, Tennessee and elsewhere trees and crops downwind of the polluting sources have withered and died.

Growers of orchids and other flowers, of tobacco and garden produce, of citrus fruits and grapes have had to give up otherwise valuable farmlands because they were too close to air polluting cities. At least half the states have suffered crop damage from smog.

Air pollution, particularly sulfur oxides, has also been tough on materials. It corrodes and tarnishes metals, weakens and fades fabrics, makes leather and rubber brittle, discolors paint and stone, etches glass and messes up electrical circuits. It has damaged priceless old masters and great sculptures.

In all the air pollution disasters, dumb animals as well as sentient human beings have died. Says the National Tuberculosis and Respiratory Disease Association in a discussion of the Poza Rica, Mexico, hydrogen sulfide spill: "All the canaries and many other birds and animals died."

One of the pollutants particularly dangerous to animals is the fluorides released into the air from factories making fertilizer, aluminum, and iron and ceramics products.

Fluorides get into plants, which concentrate them. Cattle eat the plants, Their teeth become mottled. As they feed further "on this insidious food" they lose weight, give less milk, and finally become so crippled they have to be destroyed.

Is the United States doing anything about air pollution? It is. Following the lead of California, the federal government has issued auto emission control standards—not yet as stringent as California's—for the guidance of state governments.

Under the Air Quality Act of 1967 it is creating 57 control regions which by the summer of 1970 will include all the 50 states plus the District of Columbia, Puerto Rico, and the Virgin Islands. The states and other governing bodies are responsible for enforcing the clean air standards.

According to the Senate report, however, the proposed standards won't be effective. Said the committee:

"Studies indicate that, under existing controls, automobile air pollution in the United States will more than double in the next 30 years because of the projected increase in both the number of vehicles and miles driven by each vehicle."

One trouble is that health data aren't specific enough to prove air pollution is the villain doctors feel sure it is. You can't simulate in a laboratory the precise kind of polluted air most of us breathe.

Suppose it would cost $3 billion to $4 billion a year to curb air pollution in the next 10 years. Can you justify such expenditures for health reasons alone, forgetting the esthetic, crop, animal, material damage? Not at the moment, maybe.

But the health evidence is beginning to come in. In the Department of Health, Education and Welfare has been created the National Air Pollution Control Administration headed by Dr. John T. Middleton.

Commissioner Middleton forecast recently that accumulating data will lead eventually to far more stringent controls than any yet adopted.

So far there is little reason to believe that the air will be any cleaner 10 years hence than it is now, according to federal authorities. But attempts are being made to minimize the effects of bad air.

In Los Angeles, local broadcasting stations and newspapers for some time have included smog forecasts along with the daily weather report.

Now the U.S. Weather Bureau is starting special programs in St. Louis, Chicago, New York, Philadelphia, and Washington "in support of air pollution agencies." The special forecasts will warn of weather conditions which might aggravate pollution.

This will permit local officials to "take measures to control the amount of pollutants emitted from open dumps, smokestacks, and other pollution sources." The new Nimbus 3 weather satellite is equipped to measure pollutants as it looks down through the atmosphere from space.

Such measures may help to prevent disastrous "episodes." But it remains to be seen what effect they will have on the chronic smog which, according to the federal government, "is a serious threat to public health and welfare."

Not everybody agrees with those who say the automobile is a threat to mankind. According to auto industry spokesmen, emission control devices on new cars will, by the time used cars have left the highways, return the air to the relatively clean state of 1940.

The Senate report on auto emissions called for development of steam cars to reduce air pollution. Others have urged return of the smogless battery-driven cars of old.

Proponents of nuclear power contend they would cut down heavily on the amount of pollutants now dumped into the atmosphere by coal-fired power plants. Others are worried about radioactive contaminants that might be loosed if a nuclear plant failed.

If proliferating man cannot get all the energy he needs except by burning up his fossil legacies of coal and oil or by risking pollution by radioactivity, perhaps he should fall back on the original source, the sun.

Attempts to trap sunlight economically for power have not yet succeeded on any large scale. But Dr. Cole of Cornell feels the effort should continue.

Solar radiation, he has said, must become man's chief energy source—if the species lasts long enough to see this happen. In this connection, the National Geographic Society said recently that "If man could collect and effficiently use it, the sunlight falling on just the city of Los Angeles would supply more energy than is consumed in all the homes on earth."

It will be a long time before this comes to pass. Meanwhile, where are we? The National Tuberculosis and Respiratory Disease Association in its "Pollution Primer" says:

"Nature is fighting a losing battle with man-made air pollution .. vast expanses of countryside smolder and stink. Dreamy fogs are accomplices to murder. Sunny, windless days carry, like a disease, the threat of suffocation."

Why should we care? The Environmental Pollution Panel of the President's Science Advisory Committee said this: "Man is but one species living in a world with numerous others; he depends on many of these others not only for his comfort and enjoyment but for his life."

Says climatologist Reid A. Bryson of Wisconsin: "We would like our grandchildren to experience blue skies more often. .. "

The Tuberculosis Association summed it up:

"Air pollution threatens not only man's wallet and his health. Air pollution erodes his soul. Every mountain blacked out by pollution, every flower withered by smog, every sweet-smelling countryside poisoned by foul odors destroys a bit of man's union with nature and leaves his spirit diminished by loss."

Perhaps the simplest summation of all was uttered by Russell E. Train, president of the Conservation Foundation, who became Undersecretary of Interior in the Nixon administration.

"The real stake," he said, "is man's own survival—in a world worth living in."

The Noisy Animal, Man: Fruits of Progress Can Be Bitter as well as Sweet

The racket may not be killing you. On the other hand, maybe it is—slowly, insidiously.

The din that assaults our ears almost non-stop is shattering our tranquillity, hurting our health, and contaminating an environment already poisoned by air, water and soil pollution.

Pollution of our increasingly despoiled living space by noise is another example of the now widely recognized truth that the fruits of technology can be bitter as well as sweet.

Noise pollution, which has been called "The price of progress," is getting worse every year. Nothing very effective is being done about it— in this country, at least. According to Sen. Mark O. Hatfield, R-Ore., the United States is the noisiest of modern societies.

The consensus at an American Medical Association congress in Chicago was that noise is as great a hazard to mankind as air and water pollution. It does both physical and psychological damage.

One of the speakers at the AMA congress said "The public must be made aware that offensive noise can be controlled and must be made angry enough to do something about it."

Noise has been defined in many ways. It is unwanted sound, sound without value, unrestricted sound, sound that hurts, harms, distracts, destroys sleep, invades privacy, frightens, irritates, or simply annoys.

Just how dangerous is the clatter and clamor and the mechanical screeching and screaming which assails us all and from which there appears to be no escape, not even in suburbia?

Dr. Vern O. Knudsen, Chancellor Emeritus of the University of California and a distinguished student of acoustics, the science of sound, has given this answer: "Noise, like smog, is a slow agent of death. If it continues for the next 30 years as it has for the past 30, it could become lethal."

Whether our environmental noisiness actually can kill us is debatable. But the Federal Council for Science and Technology, a White House agency, notes that "growing numbers of researchers fear that the dangerous and hazardous effects of intense noise on human health are seriously underestimated."

There is no doubt that industrial din has inflicted loss of hearing on millions of workers. At least a million workers now living suffer from some degree of deafness. The Federal Council estimates that another 6 to 16 million are exposed to noise levels which may ruin their hearing in the future.

Deafness, in fact, has finally been recognized as an occupational hazard in a lot of major industries.

But more alarming than industrial racket, because millions more persons are affected, is the steeply rising level of "community noise" which afflicts everybody—in homes, offices, schools, hospitals, even vacation resorts.

Some authorities believe that noise, along with crowding, has triggered calamities that might not have happened without the spur of noise.

Ailments which may have been caused or at least aggravated by noise include ulcers, heart diseases, allergies, and mental illness. Foreign reports have even attributed sexual impotency to high noise levels in factories.

Racket can be dangerous in indirect ways. For example: when it drowns out alarm signals or shouted warnings or verbal instructions vital to safety.

It can distract attention and interfere with the vigilance of persons responsible for monitoring controls and reacting instantly to danger signs.

A startling noise, such as a sonic boom, conceivably could cause a surgeon's knife to slip.

It has seriously been proposed that noise, as a legacy of the presumably uncaring technological "establishment," has been implicated in some way in ghetto and campus rioting.

Paul N. Borsky of the Columbia University School of Public Health says that if a person feels the noise makers are concerned about his welfare and are trying to muffle the din, he is likely to remain tolerant.

If he feels, however, that the noise propagators are callously ignoring his needs and concerns, he is more likely to be hostile. . . .

"This feeling of alienation, of being ignored and abused, is also the root cause of many other human annoyance reactions.

"This is one of the major reasons cited for urban riots, discontent by minority groups, and more recently of student revolts."

A similar idea has been expressed by Joseph J. Soporowski Jr., environmental scientist of Rutgers College. He says:

"Each of us can perhaps recall responding with indignation to assaults upon our freedoms. Yet, until recently, we have failed to respond similarly to equal assaults upon the delicate mechanism of our ears. Perhaps the cause of some of our problems and differences could be traced to irritating noise."

Warning of the "dangerous din to come," Soporowski said "Little is being done to curb this potential menace; little is being done to halt pollution of our environment by noise."

U.S. Lags Far Behind

"The overall loudness of environmental noise is doubling every 10 years in pace with our social and industrial progress."

This is the conclusion of the Federal Council for Science and Technology, one of the many groups looking into damaging effects of noise in our modern society.

"Immediate and serious attention must be given to the control of this mushrooming problem," says the council, because otherwise "The cost of alleviating it in future years will be insurmountable."

What about the charge that "little is being done to curb this potential menace?"

"There is no doubt," says the council, "that recognition of the noise problem in America has arrived late. With the exception of aircraft noise, the United States is far behind many countries in noise prevention and control."

It took the Donora smog disaster to awaken Americans to the horrors of air pollution. In October, 1948, a poisonous pall enveloped Donora, a town of 14,000 in the heavily industrialized Monongahela River Valley of Pennsylvania. Some 6,000 Donorans were sickened and 20 died.

Will it take another Donora to alert us to the dangers of noise pollution?

The Nation's first national conference on air pollution was held in Washington in 1958—10 years after the Donora tragedy. In June a year ago the first national conference on "noise as a public health hazard" was held, also in Washington. U.S. Surgeon General William H. Stewart noted that "we haven't had our Donora episode in the noise field."

"Perhaps," Stewart continued, "we never will. More likely, our Donora incidents are occurring day by day, in communities across the nation—not in terms of 20 deaths specifically attributable to a surfeit of noise, but in terms of many more than 20 ulcers, cardiovascular problems, psychoses, and neuroses for which the noises of 20th century living are a major contributory cause."

Much remains to be nailed down about the multitude of ways in which noise hurts our health and efficiency and serenity.

But, asks the Surgeon General, "must we wait until we prove every link in the chain of causation?"

"In protecting health," Stewart said, answering his own question, "absolute proof comes late. To wait for it is to invite disaster or to prolong suffering unnecessarily."

As noted earlier, noise has been called "the price of progress," the technological progress that has given us so many of the things we value, from air conditioning to vacuum cleaners. It also has given us some things we loathe, such as the cheaply built apartment houses which sound specialist Leo L. Beranek of the Massachusetts Institute of Technology calls "acoustical torture chambers."

According to the Federal Council for Science and Technology, old-fashioned dwellings of 40 to 50 years ago "were comparatively quiet places in which to live." Thanks to modern construction techniques we have "some of the noisiest buildings in existence."

Rep. Theodore R. Kupferman, R-N.Y., says New York City is busily building "the noise slums of the future."

Noise, of course, is not new. In the London of 1800 the quacking of fowl and the bellowing of animals being driven through the streets to slaughter made life hideous for people already fed up with the ceaseless clip-clop of horses' hooves on cobblestones.

Now, thanks again to technological progress, we have unmuffled scooters, motorbikes, sports cars, trucks; jet aircraft, sonic booms, screeching tires, sirens, jackhammers, air compressors; a host of kitchen and other household contrivances that whine, clank, gurgle, or rattle.

We have also clattering typewriters and cackling secretaries, the neighbors' radio and television sets, the startling crescendo of suddenly turning-on building air conditioners; the endless ringing of telephones, the din of the pile drivers, bulldozers, power saws, lawn mowers; the conversation-killing sound of dull music piped into elevators or restaurants, the distracting bells of ice cream wagons, the giant insect hum of the upstairs tenant's vacuum, the flushing of other people's toilets and the drawing of other people's baths.

Also painfully familiar to all are the night noises that sound like pistol shots but may only be backfires, puzzling and sleep destroying; the round-the-clock squawking of auto horns, the keening of handheld transistor radios; the bone tingling throb of the electric guitar, the cacophony of rock and roll bands, and the dawn chorus of the garbage collectors with their special brand of crash-bang basketball.

Make your own list. To do damage of one kind or another, a sound doesn't have to be loud—it just has to be unwanted. A Washington, D.C., psychiatrist once remarked that he was going to get a shotgun and slay the mocking bird that kept him awake with its tweedling in the night hours.

He may have been joking, or perhaps he needed a psychiatrist himself. But the serious fact remains that all of us are captive audiences: we are forced to listen to sounds—and not just mockingbird song—that we can't escape.

A favorite cliche among scientists of sound is that "one man's music may be another man's noise." Church chimes, if what you need is silence, can torture.

There are those who feel sure they love and must have hi-fi and rock and roll music, which generates noise far above the levels that would be forbidden in a well-run boiler shop.

A number of acoustics experts, alarmed by increasing evidence of hearing loss among the young, have suggested it might be a good public health measure to make discotheques display entrance signs reading "Caution: the noise levels inside may be hazardous to your health."

Noise Can Be Muffled—for a Price

Hearing loss caused by noise, according to the Federal Council for Science and Technology, constitutes "a major health hazard in American Industry."

Industries guilty or suspected of inflicting hearing hazards on their workers, according to the council, "include iron and steel making, motor vehicle production, textile manufacturing, paper making, metal products fabrication, printing and publishing, heavy construction, lumbering and wood products, and mechanized farming."

To this list may be added such military functions as "flight line and carrier deck operations, engine test cells and weapons firing, armor operations, and assorted repair and maintenance work."

No general, universally applicable standard of safe noise exposure exists in the United States. But Secretary of Labor George P. Shultz in May promulgated regulations under the Walsh-Healey Act for work done on federal contracts over $10,000 in value.

The regulations set an allowable level of industrial noise at 90 decibels for prolonged exposure (eight hours a day).

The decibel is a unit of sound used by scientists. The smallest sound an acute human ear can perceive is about one decibel. Ordinary breathing registers 10 on the decibel scale. Rustling leaves shoot the count up to 20. A restaurant where the decibel level is only 50 is quiet indeed. Other decibel values: conversation 60, heavy traffic (or an office with tabulating machines) 80, food blender 93, an amplified rock and roll band 138, the pain level for human ears 140, jet plane takeoff 140, space rocket liftoff 175.

Since the decibel scale is logarithmic, a noise high in the list (rocket) may be billions of times more powerful than one near the bottom (breathing).

Authorities appear to agree that most sounds of less than 40 decibels (the normal sustained noise inside a residence when the hi-fi and the blenders and disposal systems aren't going) are hardly noticeable.

But long exposure to levels above 80 decibels can damage the sensitive apparatus of the ear.

According to the Federal Council, "Traffic noise radiating from the freeways and expressways and from midtown shopping and apartment districts of our large cities probably disturbs more people than any other source of outdoor noise." (Aircraft noise is more intense but exposure is less than round-the-clock highway noise.)

And of all vehicles, "The trailer truck is perhaps the most notorious noise producer." At expressway speeds, a single truck may generate noise above 90 decibels while a long line of truck traffic may produce levels above 100.

At a noise conference in Washington a year ago it was noted that many trucks come equipped with fairly adequate mufflers. But operators destroy the mufflers under the impression that they reduce efficiency.

It costs money to suppress noise. The aircraft industry is spending millions in this attempt.

But not suppressing noise also costs money. Sen. Mark O. Hatfield, R-Ore., reported recently that the cost of noise to industry generally—in compensation, lost production, decreased efficiency—is estimated "at well over $4 billion (repeat billion) per year."

This loss is accumulated in bits and pieces and in subtle ways. The experts say 19 per cent more energy is needed to do a job in a noisy place than in quiet one.

Everybody says noise can be muffled—if we are willing to pay the price for doing it. Other countries, but not the United States, have included national sound insulation regulations in their building codes.

These nations are Austria, Belgium, Bulgaria, Canada, Czechoslovakia, Denmark, England, Finland, France, Germany, Netherlands, Norway, Scotland, Sweden, Switzerland, and the U.S.S.R.

"In the field of architectural acoustics and the control of noise in buildings," says the Federal Council, "we are far behind federally supported or implemented research in Canada, England, and Europe, and are currently behind the level of activity of Russian and Japanese theoretical, analytical, and applied research."

London, Berlin, and Paris have decreed leather or rubber rims on their garbage cans to curb one of the greatest noise annoyances of city life.

But only recently has New York begun even to experiment with plastic or paper bags as a means of eliminating the jangling danger of the traditional metal cans.

Some states are trying to cut down on highway cacophony. Connecticut plans to set up "sound traps" similar to the radar "speed traps" now in use.

Nonetheless, noise is escalating, along

with the number of physical and mental maladies attributed to it.

Coming up is the supersonic transport which, if it ever is put into transcontinental service, will smite the eardrums of up to 50 million Americans daily with this clap of doom otherwise known as the sonic boom.

There are those who feel that only city and state regulations can deal effectively with the rising tide of din. But the federal government, say others, has the responsibility for doing noise research and providing noise "guidelines" for local governments.

What is needed, says the Federal Council, is "a total national program to abate undesirable noise."

As W. H. Ferry, vice president of the Center for the Study of Democratic Institutions, Santa Barbara, has noted, noise like the sonic boom "is de-civilizing."

"No self-respecting civilization," says Ferry, "ought to have to accommodate itself to such an annoyance."

Uglification

We have been warned: our garbage, our junk, our rubble threaten to engulf us.

We have devoted much thought to pollution of the air we breathe and to the water we drink.

But we have paid little attention, comparatively, to what the nice-minded call "solid wastes."

Yet solid wastes in their myriad forms —everything from animal dung heaps and city garbage to universal litter and abandoned autos—are the worst of the polluters.

They pollute not only air and water but the landscape. They add "uglification" to the mess of horrors man has contrived for himself. They also constitute a reckless waste of irreplaceable resources.

If not a tribute, they are at least a monument to our affluence, our technological ingenuity, and our "heritage of waste" in a use-and-discard society.

As technology presents us with ever more conveniently packaged "consumer items" and as man's numbers mushroom, the rubbish pile grows ever higher. It is, in fact, growing faster than the population.

Charles C. Johnson Jr., administrator of the new Consumer Protection and Environmental Health Service of the Public Health Service, states it this way: "Growing mountains of garbage and trash threaten to bury us in our own waste products."

They already are hurting our health. They already have destroyed large areas of living space which nature had allotted to creatures of the wild. They already have spread "scenic blight" throughout the countryside. They have contributed their large bit to what Johnson says is a rapidly approaching drinking water crisis.

It used to be when the nation was young that no harm was done if you just threw away something you no longer wanted, if you just "spread around" your garbage.

But as the National Academy of Sciences has pointed out in a special report, "As the earth becomes more crowded, there is no longer an 'away.' One person's trash basket is another's living space."

According to Johnson, this country is now trying to deal with 3.5 billion tons of solid wastes every year. This includes 1.5 billion tons of animal excreta, 550 million tons of what's left over from the marketable parts of farm crops, 1.1 billion tons of mineral wastes, 110 million tons of industrial trash, and 250 million tons of household, commercial, and municipal wastes.

These figures do not include the millions of automobiles junked each year. It has been estimated that the car discard rate will reach eight million a year by 1975.

In addition to all this is the unguessable (certainly in the billions of tons) amount of annually accumulated debris from the demolition of buildings and highways to make way for new ones.

Each of us contributes on the average 5.3 pounds to the garbage man's haul of food scraps or rubbish. Every 30 seconds some one of us abandons a dead automobile on a city street or country roadside.

Abandoned cars are a familiar part of big city litter. About 120 a day are left on New York streets for the city to haul away. There are now between 10 million and 30 million junked cars lying about the country, disfiguring the landscape or congesting automobile graveyards.

The environs of main highways across the nation harbor some 17,500 junkyards populated largely by auto hulks.

Just to get rid of household, municipal, and industrial refuse costs us about $4.5 billion a year. Of all municipal costs, this is exceeded only by what we pay for schools and roads.

But 85 percent of this annual expenditure goes solely for collection, with only about 15 percent spent for ultimate disposal. According to one estimate, the United States would have to spend another $3.75 billion in the next five years to provide a suitable system of waste disposal.

As things stand, according to Charles Johnson of the Environmental Health Service, we "have not yet figured out what to do with the refuse that litters our countryside."

More than half of the nation's communities over 5,000 in population dispose of their wastes in a fashion described by the Public Health Service as "improper." Open dumping accounts for nearly 80 percent of all waste disposed of in this country. The Academy of Sciences report deplored this practice.

"Too often, refuse-disposal areas are open dumps—festering and disfiguring the landscape," the report said, "Flies, rats, and other disease-carrying pests find large quantities of food and suitable harborage in the piles of exposed refuse.

"The polluted drainage from open dumps is an additional insult to adjacent ground and surface water supplies. Characteristic foul odors, produced by decomposition, together with the smoke created by inefficient open burning, are often identifiable for miles."

Every cubic foot of garbage, it has been estimated, produces about 75,000 flies, not to mention rats, mice, mosquitoes, cockroaches, and other unlovely pests.

The great cities with their incinerators and "sanitary landfills" have progressed a little beyond the open dump disposal system. But the general "state of the art" remains about what it was 50 years ago. It has been said that the last real invention in waste disposal was the garbage can, and that the most recent improvement was putting an engine instead of a horse in front of the garbage truck.

Disposal means different things to different people. A housewife in a city apartment disposes of garbage effectively enough for her needs by shoving it down the incinerator shaft, or dumping it into a garbage can out front or in the alley.

The waste disposal chore is greatest in the cities where 70 percent of the national population dwells on 10 percent of the land. And even in the cities, the enormity of the problem is brought home to the householder only when the garbage men strike.

A few days without garbage collection make the streets unnavigable by the fastidious. Many of the nation's smaller communities have no regular collection services.

Considering the health hazards involved, it may be a wonder even the larger ones do. According to Richard D. Vaughan of the Environmental Control

Administration, garbage collectors "are engaging in one of the most dangerous occupations in existence."

Experts agree there is only one ultimate solution to the solid waste disposal dilemma. They call it "total recycling."

This is a dream of a nearly junkless society. In it, nothing would ever be thrown away; it would be used again. Our wastes then would become a national resource, a "mother lode" of valuable materials.

Automobiles, for example, would be designed either for reuse or for easy retrieval of their better parts. When automobiles had served out their useful lifetimes, the assembly-line process that produced them would be reversed.

Run backwards through the line, they would yield their most precious parts in a sort of priority system until only irreducible scrap, itself salvageable, remained.

But this is not the way auto makers design autos. Nor is it the way mercantile companies design the packages with which they lure the consuming public.

The modern steel industry no longer has to have scrap iron. And the people who package foods and everything else for the American family have no ecomic reason for caring what happens to the empty package.

City trash is a fantastic mixture. The most revolting part—garbage—is the easiest to get rid of. If man simply ignored it, taking the consequences to his eyes, nose, and health, nature would dispose of it.

But some of the stuff mixed with garbage in the trash haul is what scientists call "non-degradable." You can bury a nylon stocking in moist soil for years, and when you dig it up, there it is. Soil bacteria and other organisms which feast on garbage can't stomach such synthetic materials as nylon and plastics generally.

In a typical year Americans throw away 48 billion cans, 26 billion bottles, more than 30 million tons of paper, four million tons of plastics, and 100 million worn out tires weighing a million tons.

To simplify life for Americans caught away from home without an opener, technology provided the "snap top" beer and soft drink can. It was made of aluminum because aluminum cans are easier to make and tear open than cans of steel.

But aluminum is more resistant to corrosion than steel and hence harder for nature to reduce to rust. Another bit of commercial progress is the non-returnable glass bottle. Instead of lugging it back to the store for the deposit, you pitch it into the trash can along with the potato peelings.

The non-returnable bottle is looked upon almost as an enemy by those who preach "reuse and recycling" as the best answer to waste disposal problems. The committee on pollution of the National Academy of Sciences posed this question: "Should we tax glass bottles severely, or have federal law 'forbid' that they be not reused?"

According to *Solid Wastes Management* magazine, it costs New York State 30 cents for each bottle it picks up. This is seven times what it costs to make the bottle in the first place.

Such inorganic wastes in the trash mountains complicate the task of disposal. Plastics, brick, and concrete, the Academy of Sciences committee noted, "may endure for centuries."

The aluminum can, the throw-away bottle, and the plastic container have contributed more than they were ever worth to "landscape pollution." On a Sunday afternoon the Washington Cathedral grounds on Mt. St. Albans in Washington, D.C., is cluttered with cans.

City dwellers walking to the bus stop have to tread their way gingerly amongst the shards of glass bottles which have been flung to the sidewalks from automobiles.

There is no escaping the litter–or the conclusion that Americans are incorrigible litterers. On some highways it is hard to see the "no dumping" signs protruding from the cascades of dumped refuse.

According to *The New York Times*, school kids who went fishing in New York's Central Park pond for wildlife caught discarded auto tires, glass containers, beer cans, waterlogged magazines, and a blanket.

Prominently placed trash cans bearing legends reading "keep your city clean" have helped, but not enough. Where man goes, he leaves litter.

Those who have toiled gasping to the summit of Colorado's 14,256-foot Longs Peak have found awaiting them a refuse can. Some hardy ranger lugged it there.

His work was not altogether in vain; on an ordinary summer's day there is, indeed, much trash in the can. But there also is litter elsewhere—remnants of sandwiches, candy bar wrappers, drained milk containers just lying around or wedged amongst the rocks in this space so high above and so far removed from the normal range of the litterbug.

According to the Senate subcommittee on air and water pollution, "perhaps the greatest waste collection headache presented by packaged materials is littering along our roadways, in our parks, and along our rivers and lakes."

"There is a vast difference in costs," says the subcommittee, "between collecting a ton of cigarette wrappers placed in garbage cans and a ton thrown away carelessly."

Authorities now agree that disposal of solid wastes must be accomplished on a regional urban-suburban-rural basis. The old local community attitude of "take it somewhere else, but don't raise my taxes in the process" no longer is tolerable, according to a National Academy of Sciences study.

And as Charles C. Johnson Jr., administrator of the Consumer Protection and Environmental Health Service, notes: "Yesterday's city dump is now in today's suburb."

The traditional methods of disposing of big city wastes are incineration and landfill. But most city incinerators are inefficient burners, and they compound the sin of air pollution.

At best, they just reduce the volume of waste. Anywhere from five to 25 percent of solid wastes, depending on incineration efficiency, remain to be stowed some other way.

The "sanitary landfill" has had some notable successes—and some notable failures. The landfill has been used to create parks, recreational areas, and public gardens. Ideally, you pick a site which has no other use, bulldoze trenches in it, haul in compacted or shredded refuse, and cover the unsightly stuff daily with decent soil.

In practice this often has resulted in seepage of pollutant matter into ground water supplies, or in destroying the marsh habitats of wild things already losing too many struggles against encroachment by man.

As Johnson says, "most cities in the country are now destroying out-of-the-way areas of natural beauty, and polluting land, air, and water in an effort to get rid of mountains of refuse."

But landfill, though a solution of diminishing usefulness as sites become scarcer, has accomplished some fine things. In Los Angeles County an open pit mine was filled with wastes to provide a botanical garden of great beauty.

In Detroit solid wastes have been fashioned into an artificial slope for sledding and skiing in winter. Virginia Beach, Va., is converting an old landfill site into a soap box derby slope and outdoor theater.

There are other examples, but they cannot be endlessly repeated in the long future. Space is running out. It is becoming more and more necessary to haul city trash to increasingly distant disposal sites, on the land or in the sea.

San Francisco is studying a plan to haul garbage 300 miles by rail to desert burial grounds instead of dumping it as before into the bay. Philadelphia is putting into operation a program for transporting its refuse to abandoned mines 100 miles away.

For Philadelphia this will be cheaper by $2 a ton than incineration. There are some 8,000 abandoned mines in this country. Sooner or later they may all be filled with the junk of civilization.

There remains the sea. It has been suggested that offshore islands might be built of city wastes in the Atlantic for use as supersonic aircraft runways.

Frank R. Bowerman of Zurn Industries, writing in the *Investment Dealers' Digest*, has estimated that one runway could be built every year in shallow offshore waters with the eight million tons of solid wastes produced annually by New York City.

The University of Rhode Island is examining the possibility of burning city garbage in incinerator ships which would dump the ashes into the sea. This would solve the problem of the old garbage scow, the fruits of which too often washed back ashore, polluting the beaches.

New York scientists meanwhile made a discovery which may prove significant in the future—garbage tossed into the offshore waters attracted hordes of fish which, unhappily for fishermen, have moved elsewhere since this practice was abandoned.

The New York sanitation department dumped 500 junked automobiles in the ocean off Long Island last year to find out whether auto hulks would serve as artificial habitats and breeding grounds for fish. This idea has been recommended by students of the dead auto problem and even was promoted in a comic strip recently.

Who knows? Maybe it will save New York's offshore fisheries and at the same time provide an answer to what to do with old autos that have not yet found a suitable burying ground. Deep ocean canyons also have been recommended as storage for things no longer wanted on land.

A lot of other ideas have been suggested and many of them tried. Europe, perhaps because it is smaller and more crowded, has tried harder and gone farther than the United States toward perfecting efficient incinerators.

The Japanese have invented a new "dense compaction" process which presses mounds of ordinary refuse into blocks of a ton or more that will sink in water. At one time it was hoped such blocks could be used as building materials.

But gases generated by the organic material within appear to have made this an impractical and even dangerous technique. Dense compaction does, however, promise to be a boon for long-haul cleanup trains bearing trash from cities to distant disposal sites.

Much effort has been made to enlist the profit motive in the battle to save the environment from man's wastes. Tried many times but still found wanting is "composting," the decomposition of city and farm wastes through bacterial action for production of fertilizer and humus for agriculture.

The competition of more convenient chemical fertilizers has all but killed composting for profit. This also explains why the great cattle feedlots of the nation are generating piles of animal wastes faster than they can get rid of them.

Time was when organic material of this sort, spread over the fields, kept our soil rich and productive. Now there is little to do with it except let its odors befoul the air and its drainage pollute our waters.

About two-thirds of U.S. beef production comes from cattle feedlots. One cow produces waste equal to the sewage of 16 people. One feedlot handling 10,000 head of cattle has the same waste disposal needs of a city of 160,000 persons.

According to former Secretary of Agriculture Orville L. Freeman, the nearly three million head of cattle fed on lots in Nebraska and Iowa alone create an amount of waste equal to that produced by 49 million human beings, 11 times the population of the two states.

Farmers who once drove a manure spreader back and forth across their fields now find it cheaper and easier to scoop the fertilizer they need out of a bag than to dig it out of a dung pile.

Authorities are becoming convinced that salvaging human, industrial, animal, commercial and other wastes should be considered a means of cutting disposal costs rather than a way to make a profit.

Hempstead, N.Y., salvages heat from its refuse incinerator to run a 2,500-kilowatt electric power station and a 420,000 gallon a day desalting plant. But the net effect is to make waste disposal less costly, by no means to make it profitable.

Salvage probably is one of man's oldest occupations. Many persons now living remember that every little town once was roughly divided by the railroad tracks.

On one side were the stores and the homes and the schools and churches, on the other the junkyards where a reasonably active lad could collect a dime a week or more by turning in scavenged scraps of anything from tinfoil to babbitt metal or discarded copper wire.

But the small town junkyards are disappearing. In their place is a $3 billion a year industry of at least 2,300 large companies. This industry, however, is interested not in municipal wastes but in clean and easily identifiable commercial and industrial discards.

Residential trash is too mixed to be worth the modern junkman's attention. The most commonly salvaged solid wastes now are such things as cardboard, newspapers, steel cans, auto bodies, and a small amount of glass bottles.

Broken glass occasionally is salvaged and crushed for remelting into new bottles and jars. Some research is being done into use of smashed glass as a substitute for sand or gravel. The possibility has been raised of inventing a kind of bottle that will automatically dissolve after it has been emptied.

There are those who believe that solid waste disposal is the most serious of the pollution problems man has brought upon himself. This is partly because it creates both air and water as well as land pollution, and partly because mankind obviously is losing the race against burial under its own piling up discards.

If you don't believe this, look around you and try to think what the scene will be like 20 or 40 or 100 years hence, with the population doubled or tripled, or quadrupled, if some new solutions aren't conjured up in the meantime.

The government is trying to forestall disaster with a program for financing research into waste disposal schemes and assisting in local and regional plans for building demonstration plants incorporating modern disposal methods.

According to Charles C. Johnson Jr., director of the Consumer Protection and Environmental Health Service, the solid waste "environmental problem may well prove the most difficult and serious of all."

One difficulty is that waste disposal has been considered a local responsibility and, at the same time, a local irresponsibility. Communities adjacent to cities, the big waste producers, have tended to shrug off their problems as being beyond their own control.

So the federal government is attempting to distribute its research and demonstration grants as far as possible on a regional or interstate basis while at the same time not refusing assistance to communities with special local difficulties.

The goal, if man is not finally to be swamped by his own filth, is to get new use out of everything he has ever used. This is the gospel of "recycle and reuse." It is more than just an antidote to suffocation in garbage and litter.

It is the only ultimate answer to the fact that man is throwing away priceless resources he cannot recapture except by using them over and over again.

It is reminiscent of the old New England saying: "Use it up, wear it out, make it do."

This is the "economic" approach. There is another, the humane. Says Johnson of the Environmental Health Service, "In the inner city, accumulated garbage and trash create breeding grounds for rats, insects, and vermin, and constitute a major health problem."

"Before we can do anything effective in the deteriorating areas of our cities," Johnson says, "we have to attack the problem of solid waste disposal."

There is yet another, more universal, way of looking at what all of us in our profligate carelessness are doing to ourselves and to those who will inherit our soiled world. Again Johnson:

"We must halt the deterioration of the environment. We must make life worth living in the ghetto and in the suburbs, in the town house and in the cottage, in the city and in the country.

"We must prove that ugliness, danger, and misery do not have to be a part of the birthright of any American, wherever he may live in this land."

Deserts on the March

Paul B. Sears

Professor Sears is one of America's best-known botanists. He is a student of postglacial climate and the author of numerous interesting books and journal articles.

Dr. Sears, a scientist of broad culture, has raised the question, "Is the human race digging its own grave in North America?" The answer he gives us in his classic book *Deserts on the March* is anything but reassuring.

The author's thorough knowledge of nature, alive and growing, and of man's place in changing it, is presented in witty, graceful style which will cling to the memory long after the statistical data are forgotten. In consequence, the whole book is a series of dramatic climaxes designed to catch the reader's interest and to help make us face up to some sobering environmental realities.

Man, Maker of Wilderness

The face of earth is a graveyard, and so it has always been. To earth each living thing restores when it dies that which has been borrowed to give form and substance to its brief day in the sun. From earth, in due course, each new living being receives back again a loan of that which sustains life. What is lent by earth has been used by countless generations of plants and animals now dead and will be required by countless others in the future. The supply of an element such as phosphorous is so limited that if it were not constantly being returned to the soil, a single century would be sufficient to produce a disastrous reduction in the amount of life. No plant or animal, nor any sort of either, can establish permanent right of possession to the materials which compose its physical body.

Left to herself, nature, manages these loans and redemptions in not unkindly fashion. She maintains a balance which will permit the briefest time to elapse between burial and renewal. The turnover of material for new generations to use is steady and regular. Wind and water, those twin sextons, do their work gently. Each type of plant and animal, so far as it is fit, has its segment of activity and can bring forth its own kind to the limits of subsistence. The red rule of tooth and claw is less harsh in fact than in seeming. There is a balance in undisturbed nature between food and feeder, hunter and prey, so that the resources of the earth are never idle. Some plants or animals may seem to dominate the rest, but they do so only so long as the general balance is maintained. The whole world of living things exists as a series of communities whose order and permanence shame all but the most successful of human enterprises.

It is into such an ordered world of nature that primitive man fits as a part. A family of savage man, living by the chase and gathering wild plants, requires a space of ten to fifty square miles for subsistence. If neighbors press too closely, the tomahawk of tribal warfare offers a rude but perhaps merciful substitute for starvation. Man in such a stage takes what he can get on fairly even terms with the rest of nature. Wind and water may strike fear to his heart and even wreak disaster upon him, but on the whole their violence is tempered. The forces of nature expend themselves beneficiently upon the highly developed and well-balanced forests, grasslands, even desert. To the greatest possible extent the surface consists of mellow, absorbent soil, anchored and protected by living plants—a system buffered against the caprice of the elements, although of course subject to slow and orderly change. Bare ground left by the plow will have as much soil washed off in ten years as the unbroken prairie will lose in four thousand. Even so, soil in the prairie will be forming as fast as, or faster than it is lost.

Living in such a setting, man knows little or nothing of nature's laws, yet conforms to them with the perfection over which he has no more choice than the oaks and palms, the cats and reptiles around him. Gradually, however, and with many halting steps, man has learned enough about the immutable laws of cause and effect so that with tools, domestic animals, and crops he can speed up the processes of nature tremendously along certain lines. The rich Nile Valley can be made to support, not one, but one thousand people per square mile, as it does today. Cultures develop, cities and commerce flourish, hunger and fear dwindle as progress and the conquest of nature expand. Unhappily, nature is not so easily thwarted. The old problems of population pressure and tribal warfare appear in newer and more horrible guise, with whole nations trained for slaughter. And back of it all lies the fact that man has upset the balance under which wind and water were beneficial agents of construction, releasing them as twin demons which carve the soil from beneath his feet, to hasten the decay and burial of his handiwork.

Nature is not to be conquered save on her own terms. She is not conciliated by cleverness or industry in devising means to defeat the operation of one of her laws through the workings of another. She is a very business-like old lady, who plays no favorites. Man is welcome to out-number and dominate the other forms of life, provided he can maintain order among the relentless forces whose balanced operation he has disturbed. But this hard condition is one which, to date, he has scarcely met. His own past is full of clear and somber warnings—vanished civilizations buried, like dead flies in lacquer, beneath their own dust and mud.

For man, who fancies himself the conqueror of it, is at once the maker and the victim of the wilderness. Even the dense and hostile jungles of the tropics are often the work of his hands. The virgin forest of the tropics, as of other climes, is no thicket of scrub and thorn, but a cathedral of massive, well-spaced giant trees under whose dense canopy the alien and tangled rabble of the jungle does not thrive. Order and permanence are here—these giants bring forth young after their own kind, but only so fast as death and decay break the solid ranks of the elders. Let man clear these virgin forests, even convert them into fields, he can scarcely keep them. Nature claims them again, and her advance guards are the scrambled barriers through which man must chop his way.

In the early centuries of the present era, while the Roman Empire was cracking to pieces, the Mayas built great cities in Central America. Their huge pyramids, massive masonry, and elaborate carving are proof of capacity and leisure. They also indicate that the people who built them probably felt a sense of security, permanence, and accomplishment as solid as our own. To them the end of their world was no doubt unthinkable save as a device of priestly dialectic, or an exercise of the romantic imagination. Food there was in abundance, furnished by the maize, cacao, beans, and a host of other plants of which southern Mexico is the first home. Fields were easily cleared by girdling trees with sharp stone hatchets. You can write your name on plate glass with their little jadeite chisels. The dead trees were then, as they are today in Yucatan, destroyed by fire, and crops were planted in their ashes.

Yet by the sixth century all of this was abandoned and the Second Empire established northward in Yucatan, to last with varying fortunes until the Spanish conquest. Pyramids and stonework became the playground of the jungle, so hidden and bound beneath its knotted mesh that painful labor has been required to reveal what is below. Farther north in Yucatan, in humble villages, are the modern people, unable to read the hieroglyphs of their ancestors, and tresuring only fragments of the ancient lore which have survived by word of mouth. There persists among these people, for example, a considerable body of knowledge concerning medicinal plants, their properties and mode of use. But the power and glory of the cities is gone. In their place are only ruins and wilderness. Their world, once so certain, stable, dependable, and definite, is gone. And why?

Here, of course, is a first-rate mystery for modern skill and knowledge to unravel. The people were not exterminated, nor their cities taken over by an enemy. Plagues may cause temporary migrations, but not the permanent abandonment of established and prosperous centers. The present population to the north has its share of debilitating infections, but its ancestors were not too weak or wasted to establish the Second Empire after they left the First. Did the climate in the abandoned cities become so much more humid that the invasion of dense tropical vegetation could not be arrested, while fungous pest, insects, and diseases took increasing toll? This is hard to prove. Were the inhabitants starved out because they had no steel tools or draft animals to break the heavy sod which formed over their resting fields? Many experts think so.

Certainly the soil of the wet tropics is very different from the deep rich black soil of the prairies. Just as soaking removes salt from a dried mackerel, so the nourishing minerals are quickly removed from these soils by the abundant water. In the steaming hot climate the plant and animal materials which fall upon the ground are quickly rotted, sending gases into the air and losing much of what is left, in the pounding, soaking wash of the heavy tropical rains. Such organic material as may be present is well incinerated when the forest covering is killed and burned, as it was by the ancient Mayas, and still is by their descendants. Such a clearing will yield a heavy crop for a few seasons, by virtue of the fertilizer in the ashes and what little is left in the soil. Presently the yield must decline to the point where cultivation is no longer possible. A fresh clearing is made and the old one abandoned. Step by step the cultivation proceeds farther from the place of beginning. Whether the idle fields, forming an ever widening border about the great cities, came to be hidden beneath an armor of impenetrable turf or completely ruined by sheet erosion and puddling, is immaterial. The restoration of fertility by idleness has proved a failure even in temperate climates. It is not a matter of one, or even several, human generations, but a process of centuries. The cities of the Mayas were doomed by the very system that gave them birth. Man's conquest of nature was an illusion, however brilliant. Like China before the Manchu invaders, or Russia in the face of Napoleon, the jungle seemed to yield and recede before the Mayas, only to turn with deadly, relentless deliberation and strangle them.

So much for a striking case of failure in the New World. How about the Old—cradle of humanity? Here there are striking cases of apparent success, long continued, such as eastern China and the Nile Valley. On ther other hand are many instances of self-destruction as dramatic as that of the Mayas—for example the buried cities of the Sumerian desert. Let us examine both failure and seeming success; after we have done so, we shall realize how closely they are interwoven.

The invention of flocks and herds of domestic animals enabled man to increase and prevail throughout the great grassy and even the desert interior of the Old World. Food and wealth could be moved on the hoof. A rough and ready "cowpuncher" psychology was developed as a matter of course, combining a certain ruthless capacity for quick action along with an aversion to sustained and methodical labor, except for women. Living as these people did, in a region where water was none too abundant and pasture not always uniform, movement was necessary. Normally this was a seasonal migration—a round trip like that of the buffalo and other wild grazing animals. But from time to time the combination of events brought about complete and extensive shifts.

Where moisture was more abundant, either directly from rain, or indirectly thorough huge rivers, another invention took place. This second invention was the cultivation of certain nutritious grasses with unusually large fruits—the cereals. Probably not far from the mouth of the Yangtze River in southeastern China rice was domesticated, while at the eastern end of the Mediterranean wheat and barley were put to similar use, both in Irak (Mesopotamia) and Egypt. Along with these cereals many other plants, such as beans, clover, alfalfa, onions, and the like were grown. This invention provided food cheaply and on a hitherto unprecedented scale. Domestic animals could now be penned, using their energy to make flesh and milk instead of running it off in the continued movement for grass and water. Other animals like the cat and dog relieved man of the necessity of guarding his stored wealth against the raids of rats and robbers. Large animals like the ox and ass saved him the labor of carriage and helped in threshing and tillage. The people themselves became accustomed to methodical and prolonged labor. They devised means of storage and transport and developed commerce. Mechanical contrivances proved useful and were encouraged. On the other hand such folk were not celebrated for their aggressiveness nor for an itching foot. As they became organized and accumulated a surplus of skill and energy they developed great cities and other public works, with all adornments.

The history of early civilization can be written largely in terms of these two great inventions in living—the pastoral life of the dry interior and the settled agriculture of the well-watered regions. Their commerce, warfare, and eventual, if imperfect, combination make the western Europe of today. What of their effects upon the land?

Wherever we turn, to Asia, Europe, or Africa, we shall find the same story repeated with an almost mechanical regularity. The net productiveness of the land has been decreased. Fertility has been consumed and soil destroyed at a rate far in excess of the capacity of either man or nature to replace. The glorious achievements of civilization have been builded on borrowed capital to a scale undreamed by the most extravagant of monarchs. And unlike the bonds which statesmen so blithely issue to—and against—their own people, an obligation has piled up which cannot be repudiated by the stroke of any man's pen.

Uniformly the nomads of the interior have crowded their great ranges to the limit. We shall see later what a subtle matter this crowding may be—the fields may look as green as ever, until the inevitable drier years come along. Then the soil becomes exposed, to be blown away by wind, or washed into great flooded rivers during the infrequent, usually torrential rains. The cycle of erosion gains momentum, at times conveying wealth to the farmer downstream in the form of rich black soil, but quite as often destroying and burying his means of livelihood beneath a coat of sterile mud.

The reduction of pasture, even with the return of better years, dislocates the scheme of things for the owners of flocks and herds. Raids, mass migrations, discouraged and feeble attempts at agriculture, or, rarely, the development of irrigation and dry farming result—and history is made.

Meanwhile, in the more densely settled regions of cereal farming, population pressure demands every resource to maintain yield. So long as rich mud is brought downstream in thin layers at regular intervals, the valleys yield good returns at the expense of the continental interior. But such imperial gifts are hard to control, increasingly so as occupation and overgrazing upstream develop. In the course of events farming spreads from the valley to the upland. The forests of the upland are stripped, both for their own product and for the sake of the ground which they occupy. Growing cities need lumber, as well as food. For a time these upland forest soils of the moister region yield good crops, but gradually they too are exhausted. Imperceptibly sheet erosion moves them into the valleys, with only temporary value to the latter. Soon the rich black valley soil is overlaid by pale and unproductive material from the uplands. The latter may become an abandoned range of gullies, or in rarer cases human resourcefulness may come to the fore, and by costly engineering works combined with agronomic skill, defer the final tragedy of abandonment.

Thus have we sketched, in broad strokes to be sure, the story of man's destruction upon the face of his own Mother Earth. The story on the older continents has been a matter of millenia, as we shall see. In North America it has been a matter of not more than three centuries at most—generally a matter of decades. Mechanical invention plus exuberant vitality have accomplished the conquest of a continent with unparalleled speed, but in doing so have broken the gentle grip wherein nature holds and controls the forces which serve when restrained, destroy when unleashed.

The Wisdom of the Ages

Is the battle really a losing one to date? Have not invention, energy, and discipline consolidated the gains of mankind securely against all danger, excepting our own selfishness and capacity for mutual destruction in time of war and peace? What of the wisdom of the East? What of the vast plain of eastern China, which feeds one-quarter of the human race? What of Mother India, whose peoples have increased under British rule? What of the narrow valley of the Nile, as populous, and seemingly as fertile as ever? Are not these the reality, with man in equilibrium with nature as long as both shall last? Are not the cases of retreat and destruction mere incidents, inconclusive and temporary?

China is larger than the United States of America, and more diverse. Her people are as clever in handling the land as any. She has, in a literal sense, forgotten more than the West ever knew. To her the West is indebted for rice, the peach, the soy bean, alfalfa. From her southern boundaries came the apple, which even today forms wild mountain forests in the

Himalayas. Her emperors maintained peace and communication between her provinces for long periods. New varieties of many plants were developed and their culture spread. It is said, and probably with truth, that the land of southeastern China is almost unique in bringing forth as heavy yields today as it ever has. Her people are industrious, frugal, and intelligent. If allowed to come here and compete, they could drive our thriftiest farmers out of business. What of China?

To begin with, China contains millions of people who are never far from the verge of starvation. Nearly every year of her history witnesses some flood or famine, even though local in character. When seven thousand Chinese Communists were wiped out in four days the government reminded its critics, truly enough, that the cost in lives of this particular project in "political sanitation" was a small matter beside the deaths in a single provincial famine. Every foot of tillable land throughout most of China is terraced and contoured with geometrical perfection. Her farmers save every scrap of garbage and other organic matter. They trudge long distances to the cities and villages with containers across their shoulders, to bring back to their farms what the writers on agriculture call by the euphemistic name of "night soil"—human excrement. All of this, together with the black rich ooze retrieved from their rivers following high water, is worked into their tiny fields with more pains than a Dutch housewife bestows upon her window garden. In some way, too, the Chinese farmer has learned the value of legumes in improving the soil. Unlike the Europeans, who knew it in the days of Rome, but neglected it thereafter, he has consistently practiced legume rotation, growing among other things soy beans, from which Chinese cooks evolve an imposing array of edibles. All of this is not to say that Chinese agriculture is perfect—it has much to learn from modern science. For example, the importance of seed selection is not properly understood, and inferior varieties of plants are often grown. But the fact remains that the general standard of practice is so far above that in many parts of the Western world as to admit no comparison. One of the finest achievements of the Chinese farmer has been the conversion of the Red Basin of Szechuan from an incipient bad land, supporting less than a hundred forty-five thousand people in 1710, into a flourishing, beautifully terraced agricultural countryside of forty-five million inhabitants.

Yet nothing is more mistaken than to think that China is in any real sense self-sustaining. She is using fertility stored by the work of her great rivers during millions of years past, supplemented by present tribute of an area twice the size of China proper, but with one-sixteenth as many people. In fact her two great rivers, the Yangtze and the Hwang, rise in Tibet, with its sparse population of less than three people to a square mile. These streams are fed by seemingly inexhaustible snows, and bear mineral matter, along with vegetable material afforded by the world's largest mountains and their lush plant cover. Let modern industry penetrate to these sources and exploit them with the zeal that has been expended upon our own Rocky Mountains, and China proper is doomed.

Without doubt the production of these upper watersheds is more a matter of happy chance than of deliberate policy. If means existed for the ready transport and marketing of their timber, it doubtless would have been stripped before now. As it is, the pressure of population continually surges against these forested mountains. Some decades ago potatoes were introduced into western China by the missionary priests in the hope that they would provide an additional insurance against recurring famine. As in western Europe when first introduced, the potato was regarded by the Chinese with contempt, as an inferior food suitable only for those who could not do better by themselves. Its cultivation was not too efficiently managed. Now the potato probably originated in the Andean highlands of South America, and can thrive at higher elevations than most of the staple food crops of western China. When this fact was discovered, there was a rush to strip the forests and replace them with potato patches. Eventually—in fact very shortly—the same thing happened that took place when the Irish became too dependent upon the potato. The fields were blighted by disease. No other crop afforded reserves, and the land so recently put to use had to be abandoned. This occurred on the upper valley of the Yangtze, where climatic conditions favor the quick return of the forest. But let such efforts continue and not even the genial climate of southern China will suffice to protect the upper valley against serious erosion, and the lower Yangtze against destructive flood.

Unlike the Yangtze, the Hwang, or Yellow River, is notorious for its disastrous floods. Starting, too, in the snowy, wooded, sparsely peopled mountains of Tibet, it flows down through a drier, more continental region than its southern fellow. Much of its upper valley is plateau, not unlike our high plains of the West. Here the pressure of population has been insistent. Herds and flocks have taxed the pastures to their limit, and cultivation has been attempted wherever possible. But in such a climate there is no friendly surplus of atmospheric moisture to encourage the speedy return of vegetation, once it is destroyed. Conditions here have favored the destructive action of wind and water, doubtless long before the advent of man upon the scene. Whatever plant life there is to anchor the soil is present, as in the high plains of our own country, only by the grace of unceasing struggle throughout the ages.

In consequence much of the Hwang Valley is today a region in which raging flood alternates with blinding dust storms. The infrequent but torrential rains carve the landscape and bear their toll of yellow mud into the stream and down into the populous lowland. Drought follows, and the denuded face of the landscape becomes the plaything of the ever-blowing winds. The damage is perhaps aggravated by the fact that much of the soil was originally brought in by wind from the Gobi Desert and elsewhere, during periods of arid climate. Such soil, known as *loess,* is very abundant in Asia, covering, it is estimated, about 3 per cent of the continent. Although very fertile when supplied with water, it forms vertical cliffs when eroded, and is speedily converted into a fantastic region of bad lands, unsuitable for any use. Lying quite outside the area of true climatic desert in Asia, we thus have extensive areas of cultural, or man-made desert, not only useless in themselves, but a menace to the fertile, well-watered plains to the east, downstream.

It should be clear then that the face of China is not without its man-made scars. She is never far from the brink of starvation, holding her own in fairer spots only because of the utmost economy, hard toil, and huge reserves beneath her soil as well as beyond her immediate boundaries. And China is a nation in which prudent land management has been an official policy since 2700 B.C., the reign of the Emperor Shen-nung.

So much for China. What of India, whose teeming millions flit past the occidental imagination in a kaleidoscopic mixture of splendor and wretchedness? In montage we may see heaps of jewels, plump elephants, forests of precious woods, groves of spice and tea, rich fields of cane and rice. But always there are flashes of hungry faces, ragged bodies, crowding beggars, and all that symbolizes misery. No matter how many rotund and prosperous Indians one may have met face to face, the inescapable image of India to most Western minds is that of a gaunt and hopeless human figure, standing on sunbaked, barren clay beside an undernourished, wizened cow. Lest this seem too gross a caricature, let an Indian, the Agha Khan, speak:

"The ill-clad villagers, men, women and children, thin and weakly, and made old beyond their years by a life of underfeeding and overwork, have been astir before daybreak, and have partaken of a scanty meal consisting of some kind or other of cold porridge of course without sugar or milk. With bare and hardened feet they reach their fields and immediately begin to furrow the soil with their lean cattle of a poor and hybrid breed, usually sterile and milkless. A short rest at midday, and a handful of dried corn or beans for food, is followed by a continuance till dusk of the laborious scratching of the soil. Then the weary way homewards in the chilly evening, every member of the family shaking with malaria and fatigue. A drink of water, probably contaminated, the munching of a piece of hard black or green champatire, a little gossip around the peepul tree, then the day ends with heavy unrefreshing sleep. in dwellings so insanitary that no decent European farmer would house his cattle in them."

The specter of famine is never far away in India. It is estimated that in the famine of 1770 ten million died, amid scenes of suffering so harrowing that only a morbid mind would dwell upon them. So slight was the margin of supply that even three successive years of good crops thereafter could not restore the balance. Not enough people were left to work the fields. In 1865 one-third of the population of Orissa died by famine, and in the subsequent three years one and half million victims were claimed. The roster of Indian famines reads with an appalling monotony.

The Indian Empire is smaller than Greater China, but compares roughly with China proper in area and population. Northward the humid Gangetic plain and the arid Punjab are watered, as in China, by the huge Himalayas. Hence they are mostly covered by rich material washed over their surface. In the same sense as China, they are dependent upon treasures stored up in the past, as well as upon water and fertility brought today from the thinly populated Tibetan regions beyond their borders. They enjoy the added protection which comes from very moderate agricultural exploitation of their northern boundaries. Here considerable areas are occupied by tribesmen whose farming operations have been of a casual nature. More or less on the move, they have allowed their fields to grow back into forests after a few years of use—a custom which the climate permits. Thus the protecting girdle of vegetation is maintained from Tibet south into the hills of India. Should this be methodically stripped and the land put to the plow, it is easy to see that great skill would be required to prevent disaster to the more populous, lower valley regions.

The Dekkan, or peninsula of India, lying south of the regions named, is separated from them by a line of low mountains from which it receives the drainage. With the exception of its west coast, it is immediately dependent upon the rainfall, yet supports a dense population. The soil is peculiarly subject to erosion, and this evil has been intensified by removal of the forest cover from the hills which form the watersheds. Modern industrialism has encouraged and hastened this process. While the soil is varied, the direct cause of famines has been largely a recurring lack of rainfall. Soil management has not been on a level with that in China. Until recently fertility seems to have been maintained by a process of growing very little more than was needed to sustain life, and consuming it locally. In the absence of means of ready transportation even a local crop failure could produce starvation, and often did just that. Today, although British rule has alleviated conditions and permitted a vast increase in population, there is still hunger in case of local crop failures. Food in ample quantities can be brought in, but no one has the means to pay for it when it comes. Furthermore, the very railroads which facilitate the transportation of food have hastened the clearing of the forests which had earlier helped to check erosion.

Great Britain has not been indifferent to the problem of soil management and food production in India. In the drier parts there had already been a thrifty, well-managed irrigation economy, which she has undertaken to encourage. British agricultural experts are convinced that India can, under proper agricultural management, raise much more than sufficient for her own needs. But in the same breath they talk seriously about maintaining the soil in good condition if that is done. Production cannot be speeded up without recourse to artificial fertilizers and newer methods to which the population is not trained. The cattle of India are largely of inferior stock and undernourished because of food shortage during the dry season. In many places so great is the pressure for fuel that dung is gathered for that purpose, and so is not returned to the soil as it should be. In the newly opened irrigated regions where water is now abundant and the soil naturally good, increasing care will be required to keep the latter in proper condition in the face of oft-repeated waterings.

As to India, then, there is no warrant for talking of a really self-sustaining community. The richer parts are so because of what they receive from beyond their own borders. As to the rest, the price of survival has been a state of suspended animation for the whole populace, low vitality, and the constant threat of hunger. What has the upper hand here, man or nature?

But let us move on toward Egypt, that half-oriental teacher of the Western world. For practical purposes Egypt consists of a narrow thread of farm land along the lower Nile, widening into a delta as the mouth is approached. On her eight million acres are supported about ten million people. That is a population twice as dense as the most populous provinces of China or India. Her people are mostly hard-working, placid, peasant farmers, little concerned about sweetness and light, or the higher life. They use primitive tools, and methods that are age-old. In addition to the staples of ancient days, they now produce an excellent grade of cotton, in increasing amounts since the American Civil War. Heretofore there has been for them little trouble about keeping their soil fertile. The annual bath of mud, administered with gentle dignity by the stately Nile, has taken care of that. Along with the mud came water which could be impounded and fed out as needed by means of reservoirs in the upper part of the valley. In the delta, however, where there are no high levels in which to store water, only one crop could be grown, while the soil was still moist from the flood. Knowing that the climate would permit crops at any time of year, if only water were available, the authorities, in true modern high-pressure style, worked out a plan of ditches through the delta, to hold water in storage. By this means three crops a year could be grown, and the ground kept constantly at work. Great stuff!

Too good to last, in fact. The ditches kept the flood from anointing the fields with mud as usual, and soon the soil of the delta of the great river Nile, richest in the world, began to show symptoms very much like those of a broken-down, one-crop cotton or tobacco farm. It is clear now that if this soil is forced, it will exact compensation in the way of fertilizers.

And as to man's ever being self-contained and self-sustaining within the Nile Valley, not even the demagogues of old Egypt would have dared preach that nonsense. The entire vast watershed of the river renders its tribute of water and topsoil to keep Egypt alive. In all the world today there is no better example of adroit, peaceful, absolute control of one nation over another than Great Britain exerts over Egypt by her possession of the Sudan, through which the river flows on its way to Egypt. She can throttle down the lower country to any degree she wills by withdrawing water to irrigate her own land. When a British high official was murdered in Cairo by a political assassin, the injured power countered by increasing the area of land under cultivation in the Sudan. For years Great Britain has been devoting some of her best scientific skill to thorough study of agricultural possibilities in the Sudan. Whether this has been primarily with a view towards producing more cotton and other raw materials, or to keep the Egyptians in a properly thoughtful frame of mind, is a topic upon which we need not expect His Majesty's government to be garrulous. Either way you take it you have a triumph of statecraft.

Outside of Egypt, what of the rest of northern Africa? The seacoast, once a famous wine district, is now unfit for that purpose. Egypt herself, beyond the immediate borders of the Nile, is desert. Both north and south of the Sudan is every evidence of misuse and deterioration—ground cover gone, soil washed and blown. In East Africa the pastures have been so overloaded with stock that here too the soil has been exposed and washed away. A curious exception occurs in places where the infestation of the accursed tsetse fly has kept down the cattle population. This pest, which has been a source of so much loss and destruction to the cattle industry, actually appears as a blessing when the long-time welfare of the continent is considered. It reminds us of the potato blight in western China, destroying food, it is true, but preventing destructive agriculture which would damage the watersheds supplying the populous districts near the coast. Again, it is like the boll weevil in the southern states, which, by ruining the cotton crop, finally forces the farmers to diversify their crops, to their own lasting benefit. Or like the wild hill people who occupy the northern reaches of India with their casual, precarious agriculture—insufficient to give them more than a meager living, but allowing the forest to keep its foothold so that the main part of India does not suffer as it otherwise would.

There is not much in the story of China, India, and Egypt to suggest that an entire continent can be exploited with the efficiency of the machine age, while its inhabitants multiply and enjoy what the politicians speak of as the "American standard of living."

13.

Dwindling Lakes

Arthur D. Hasler and Bruce Ingersoll

Dr. Hasler is currently Professor of Zoology and Director of Limnology at the University of Wisconsin. He is an eminent ecologist whose field of specialization is in the area of eutrophication, the process of lake aging.

Bruce Ingersoll, his coauthor, has been a lumberjack in the forests of Washington's Olympic Peninsula, a guide for fishermen in Wyoming and Minnesota, and a reporter for the Chicago Tribune. He is now working in the field of conservation communication at the University of Wisconsin.

Algae, minute and one-celled, ride the summer waves of every waterway in America. They colonize stagnant sloughs of rivers and gather in backwaters behind dams. Under a microscope, these tiny and rather attractive plants appear innocent enough, yet they can quickly cover a bay with scum, form hairy filaments to enslime a rocky shoreline, or clog an entire lake. With life-spans of only a few days, algae can make any lake grow old thousands of years before its time.

To realize their immense potential for harm, however, algae depend on man. It takes man to speed up natural eutrophication, the normal process of enrichment and aging undergone by bodies of fresh water. By fertilizing the nation's waters with nutrients vital to algae growth and reproduction, primarily nitrogen and phosphorous, we turn eutrophication into an accelerated, cultural process—cultural in that we are perverting nature with municipal sewage, industrial wastes, agricultural drainage, and other odious byproducts of our civilization. Cultural eutrophication, therefore, is an aberration: a natural process running amok.

In enriching the water with the nutrients in sewage, groundwater, and urban and rural runoff, we promote the exponential reproduction of algae: the seemingly harmless alga becomes 2 algae, then 4, 16, 256—multiplying until there are billions. And during such a population explosion, lakes become murky and fetid under the August sun, while wave-tossed weeds, bloated fish, and dead algae rot in shoreline windrows.

More than one-third of America's 100,000 lakes are showing signs of cultural eutrophication. The danger of accelerated eutrophication continues to grow as our population makes greater demands on national water resources, and lakes continue to take on the eutrophication syndrome. Besides showing the symptoms of excess nutrients and algae bloom, they are characterized by the depletion of dissolved oxygen in deeper waters, a change from cold-water game fish to "rough" bottom feeders, and the encroachment of rooted vegetation from shore.

The appearance of these symptoms follows a definite sequence. First, the algae population skyrockets in a man-fertilized lake. Water fleas (miniature freshwater shrimp), the staple in the diet of fry and minnows, cannot eat enough algae to keep these plants in check. As a result, billions of algae live their languid lives, reproduce, and then die. As they drift toward the bottom, their decomposing bodies exhaust the deepwater oxygen supply. Trout, whitefish, and other fish species suffocate in the oxygen-thin depths.

The lake's ecology, initially upset by excess nutrients, then becomes totally upended, since bacteria can convert only some of the dead algae into plant and animal food. Therefore, generation after generation of algae settle on the bottom, adding layer after layer to the muck. The rate of sedimentation is most rapid in a northern lake where bacteria grow only during the summer while nutrients are added throughout the year. As erosion and sedimentation fill the lake, shoreline vegetation impinges on the open water. In time, the lake becomes so shallow and overgrown that it becomes a marsh or bog. The accelerated process of aging has taken its toll: the lake's life is ended.

Cultural eutrophication is not new, nor is it solely an American problem. Recent core samples of an Italian lake indicate that ancient Roman road builders caused eutrophication by exposing nearby nutrient-rich limestone strata to erosion.

By the latter part of the nineteenth century, a few scientists in Europe and the United States, alarmed about cultural eutrophication, issued warnings that went unheeded. In 1896, on Switzerland's Lake Zürich, the problem finally received notice, and during the 1920's and 30's worldwide scientific interest began to focus on the problem. But research without implementation is impotent, and by the 1940's eutrophication was no longer confined primarily to farm belts and urban areas. It had followed urban man in his quest for recreation into the wilds.

It took the visual (and olfactory) impact of a huge body of water, Lake Erie, suffocating as a sump for industrial waste, sewage, and urban and rural run-off to bring the problem of water pollution dramatically to the public eye. Some now pronounce Lake Erie "dead." We, however, extend hope for recovery. For if communities and industries in Lake Erie's drainage basin cease polluting and fertilizing its waters, and if a cleansing flow from Lake Huron is permitted to reach and "flush out" the lake via the Detroit River, Erie might show signs of recovery within a decade. True, our hope is contingent upon many "ifs." However, it cannot be too strongly emphasized that these situations are reversible—that bodies of water will respond when the right steps are taken.

There is reason to fear for the other Great Lakes, together the greatest reservoir of clean fresh water in the world. Although Lake Superior and Lake Huron bear only a few signs of eutrophication so far, alarming amounts of DDT and other persistent toxic pesticides have been detected in the highly prized flesh of lake trout caught in their waters.

Lake Michigan is also deteriorating rapidly from increasing effluents and may soon go the way of Erie. The stench of algae and weeds decaying on beaches, compounded by the waveborne plague of dead alewives, makes eutrophication impossible to ignore in this once pure lake. Such cities as Muskegon, Gary, Chicago, Milwaukee, and Green Bay realize that the fate of Lake Michigan—to which they once owed their existence—is now uncertain.

Stewart L. Udall said a year ago that there is still time to save Lake Michigan, but warned that further delay in action would prove fatal. His statement is admittedly, and of necessity, vague, but it is correct.

The long-range outlook here is not encouraging because the mitten-shaped lake isn't in the mainstream of Great Lakes water circulation, and thus receives very little cleansing flow. Its tributaries carry heavy nutrient loads to

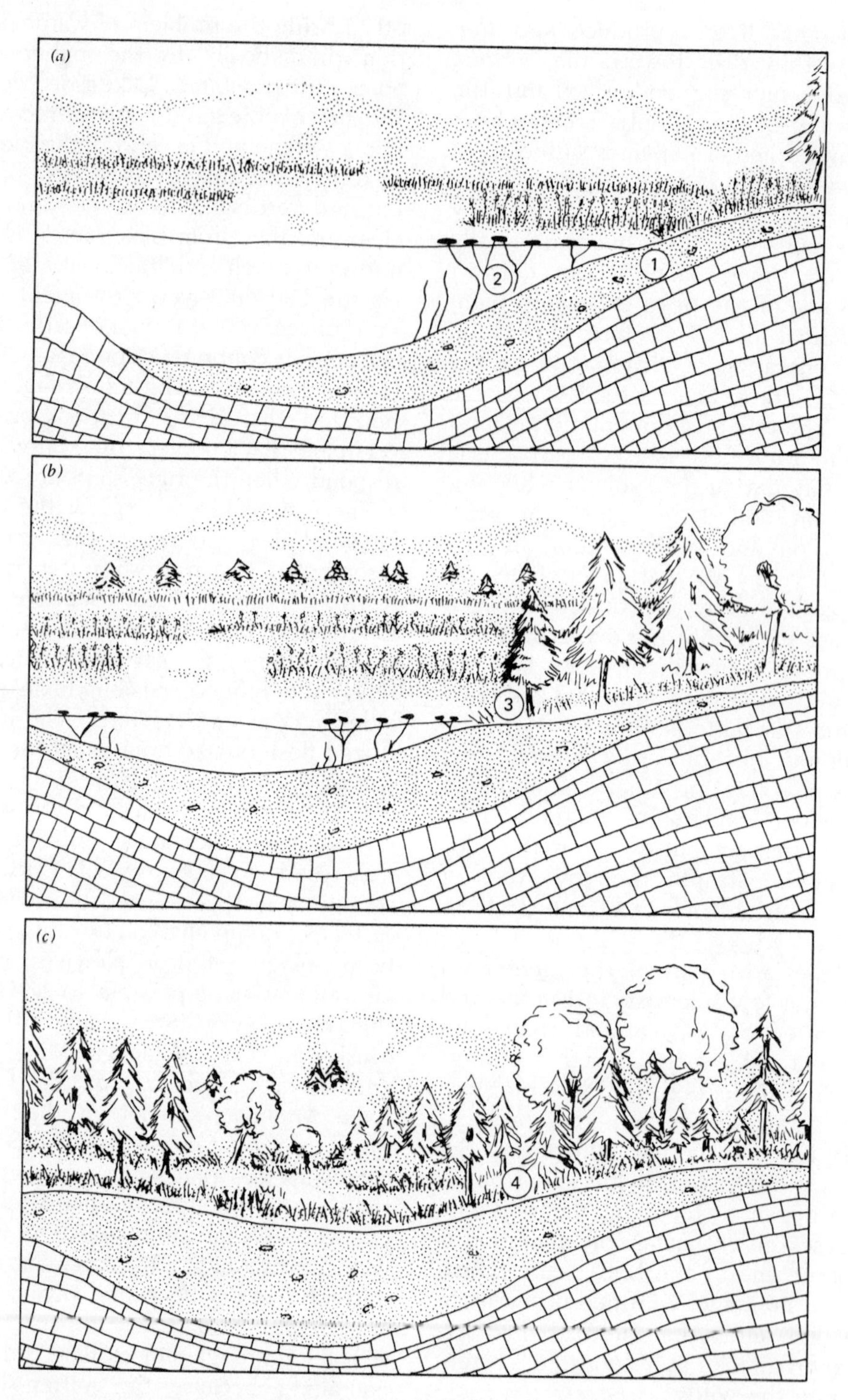

Stages of Natural Lake Succession. Figures (A to C) portray the natural aging and disappearance of a lake in the Northern Hemisphere. A newly formed lake contains few algae or eroded materials. Later, shoreline runoff brings eroded soil (1) and plant nutrients into the lake; aquatic vegetation (2) (including algae) flourish and die, adding to debris on lake bottom; marsh plants and water-tolerant conifers (3) grow at lakeside. Eventually, the entire lake is filled and covered with forest vegetation (4).

its shallows, and consequently, among Michigan's southern rim scientists have found 44 times more phosphorus (in the form of sewage effluent and industrial waste) than the lake can handle.

Lake Michigan's fate would have been decided long ago had Chicago's engineers and city fathers lacked the foresight and imagination to reverse the lakeward flow of the Chicago River and channel it into the Illinois River, a tributary of the Mississippi. The metropolis now draws water from the lake, and discharges sewage effluent into the altered river system. As is sometimes the case in such diversion operations, they have alleviated their own problem but have contributed to the water problems of St. Louis and other cities located farther down the Mississippi.

While concern for the future of the Great Lakes intensifies, distress is being voiced over eutrophication on many smaller bodies of water. Fishermen complain that dense mats of algae and rooted weeds make trolling for pike impossible, and that the few fish they do manage to catch are too tainted to eat. A summer cottager realizes that he will be landlocked if his bay becomes any shallower, and a concerned parent forbids her youngsters to wade in the green slurry of algae.

To stop cultural eutrophication, the sources of nutrients for algae growth and reproduction must be pinpointed. A 1967 survey, representative of the Midwest, indicates that 6 per cent of the nitrogen and 2 per cent of the phosphorus reaching Wisconsin waters come from septic tank seepage. Twenty-five per cent of the nitrogen and 56 per cent of the phosphorus come from municipal sewage treatment plants. Runoff from manured fields accounts for 10 and 22 per cent of these elements, while urban runoff supplies 6 and 10 per cent. Groundwater and direct rainfall on Wisconsin bodies of water contribute, respectively, 42 per cent and 9 per cent of the nitrogen, but together less than 4 per cent of the phosphorus. Industrial wastes which bypass municipal treat-

Lake	*Location*	*Excellent*	*Good*	*Fair*	*Poor*	*Endangered*	*Improving*
Crater	Oregon	■				■	
Superior	Great Lakes	■				■	
Tahoe	Calif.–Nev.	■				■	
George	New York		■			■	
Cayuga	New York		■			■	
Washington	Washington			■			■
Okoboji	Iowa		■				■
Mendota	Wisconsin			■		■	
Erie	Great Lakes				■	■	
Douglas	Michigan		■			■	
Apopka	Florida				■		■
Okeechobee	Florida		■			■	

Condition of twelve United States lakes.

ment are a source of trace amounts of each of these chemical substances.

The nitrogen contribution of groundwater and rainfall will continue to be significant as long as automobile exhausts and industrial smokestacks keep spewing nitrogen into the atmosphere. Rain, in cleansing the air, picks up this nitrogen and deposits it directly into lakes and streams or indirectly in groundwater. Nitrogen-laden rain falling on land also percolates down to the subterranean water table, the level of groundwater.

It must be remembered that the natural rate of eutrophication is the base rate for cultural eutrophication. Moreover, the natural rate depends on whether the soil in a lake's drainage basin is rich or poor in nutrients. For this reason, thousands of lakes in a nutrient-poor belt stretching from northern Minnesota and Ontario to Maine and Quebec have changed very little since the last glacier retreated 10,000 years ago. Found in sandy or granitic terrain and unspoiled by man, these deep lakes have kept their gin-clear purity and their youth—they are oligotrophic. However, should we disturb their basins and fertilize their waters, they would immediately undergo cultural eutrophication.

Seepage from the septic tanks of just a few summer cabins and resorts, for example, is rich enough in nutrients to speed up the aging process. Cochran Lake, once a pure gem set in the northern Wisconsin wilds, has deteriorated so rapidly since the first of seven cottages was built on its shores ten years ago that it now looks like a 300-acre caldron of pea soup.

Lakes with large drainage basins in limestone terrain, which is usually high in phosphorus, are far more likely to show their age than lakes in granitic or sandy basins. Because of their high nutrient content, they become shallow and die under encroaching cattails, reeds, and marsh grasses. But even though such lakes are already highly fertilized through natural means, whenever man is present his actions become the determining factor in the rate at which lakes fill in. Unless septic tank seepage is stanched, Cochran Lake—and thousands of other lakes in the resort areas of northern United States, Canada, Scandinavia, and the Alps—will be ruined.

While faulty septic tanks are a major source of nutrients in rural regions, their over-all contribution does not approach that of municipal sewage treatment plants. It has been estimated that 260 million pounds of phosphorus and 511 million pounds of nitrogen reach the nation's surface waters in the form of municipal sewage each year. Substantial amounts of these plant nutrients are discharged in effluent even after sewage is treated. As much as 75 per cent of the phosphorus in sewage comes from detergents. In addition to phosphorus and nitrogen, vitamins, amino acids, and growth hormones have been found in effluent—substances which contribute to the growth of algae and weeds. These growth stimulants are synthesized inadvertently in the biological processes of sewage treatment.

Since cities and villages across the country are rather impartial in dumping their sewage, rivers too are receiving their share of nutrients for eutrophication. This is most obvious wherever rivers have been dammed and currents slowed enough to give algae a chance to multiply. Because most of the nation's rivers have been systematically converted into series of impoundments since the 1930's, cultural eutrophication in these man-made lakes is now a common problem.

Algae blooms foul many of the man-made lakes in the TVA system, as well as the backwaters of hydroelectric dams on the Missouri and Snake rivers. The Mississippi—"father of waters"—carries partially treated sewage from Minneapolis and St. Paul forty miles downstream to Lake Pepin where algae thrive on sewage nutrients. As algae die and settle to the bottom with other sediments, the sluggish lake becomes shallower, increasing the possibility of spring flooding at Winona, Minn., and other downriver towns.

Formidable as it is, cultural eutrophication isn't inevitable. For example, five years ago officials from 70-plus local government bodies became alarmed about algal slime at sewage outlets on Lake Tahoe on the California-Nevada border, touted as one of the clearest lakes of North America. With the entire Tahoe basin's livelihood—tourism and gambling—at stake, they banded together and consulted sanitary engineers and limnologists. The prescribed cure: sewage treatment and diversion.

The South Tahoe Public Utility District, spurred by action committees of panicked businessmen, obtained a $10 million grant from Washington, raised another $9 million, and this year built a plant to treat sewage and a pipeline to carry effluent over a mountain pass to an irrigation reservoir. In diverting the effluent from the lake basin, the South Tahoe district provided California desert farms with nutrient-rich water. Meanwhile, the 42 voters of Round Hill, Nev., (most being gaming-house owners) floated a $5.8 million bond for their own sewage treatment-diversion project, fearing that smelly scum would hurt business at their lakeshore casinos and hotels. And if other Tahoe communities act right now, eutrophication should never smirch the lake's reputation for clarity—at least not for several thousands of years.

Diversion is currently the most effective measure we have to combat eutrophication, since it immediately deprives nuisance vegetation of nitrogen, phosphorus, and other nutrients. Moreover, until some technogical breakthrough occurs in sewage treatment—now far from a completely effective process—diversion will continue to be the logical approach.

Sanitary engineers haven't yet found a process, chemical or biochemical, that offers an economical alternative to diverting effluent around lakes. Nevertheless, research is being pressed throughout the country. A pilot plant, for example, has been built to test a chemical process at Ely, Minn., the launching point for canoe trips into the Superior National Forest. Research teams have investigated the feasibility of using algae to remove nutrients from effluent. The plan is to grow algae in effluent-filled ponds until they have consumed all the nutrients. Then the algae would be harvested and the nutrient-poor effluent discharged directly into a lake, obviating the need for diversion. The researchers, in practice, have been stymied by several technical problems: for one, how can algae be put to use? Unfortunately, there is no demand for algae as a fertilizer or as an animal food.

While the logic of diverting effluent is obvious, particularly in light of the shortcomings of sewage treatment, it often hasn't been accepted. Opponents of diversion in Madison, Wis., for instance, argued that runoff from manured farmland—not municipal sewage—was the chief cause of eutrophication in Lake Monona, one of four lakes linked by the sluggish Yahara River. Nevertheless, soon after Madison began diverting all of its effluent around Lake Monona in 1936, the amount of copper sulfate needed to kill algae dropped from carloads to a few handfuls by 1956.

As Madison learned in the late 1930's, diversion requires a sense of responsibility and a regard for downstream communities. That city, in sparing Lake Monona, poured its effluent into Lake Waubesa and indirectly into Lake

Kegonsa. This half-measure promptly fertilized their waters. Residents, enraged by the proliferating algae that threatened to ruin their lakes, sued the city. In 1941 legislators passed an antipollution bill aimed at Madison. Gov. Julius Heil, however, vetoed it on the ground that there was no consensus on the main cause of eutrophication. Although the bill was later enacted, it was not until 1959 that Madison's effluent was diverted completely from Waubesa and Kegonsa.

Foes of diverting sewage plant effluent around Seattle's Lake Washington to Puget Sound claimed that farm runoff was responsible for cultural europhication—just as had been claimed in Madison—even though the 24-mile long lake was absorbing the effluent of ten treatment plants by 1950. In this case, it took the lavender bloom of *Oscillatoria rubescens.* the same alga that discolored Lake Zürich, to awaken the public. Mass apathy dissolved, but a welter of conflicting opinions took its place and for a time threatened to defeat the attempts to divert sewage around Lake Washington. Eventually, 5,000 women, backed by the Municipal League and the League of Women Voters, went from door to door, mustering support for a bond issue to be used for diversion. Recently, voters reversed the stand they took in 1956 and approved the bond issue and a charter for a new metropolitan governmental body that would cope with the sewage problem. Although the diversion project hasn't been completed, the picturesque lake is nearly as clear as it was eighteen years ago. Its clarity should continue to improve as its effluent is diverted to the tide-washed sound.

Diversion projects are sometimes condemned as costly boondoggles, since rainfall inevitably contributes so much nitrogen to lakes, whether or not diversion is practiced. Critics, however, fail to consider, or choose to ignore, the fact that municipal sewage—not rainfall—is the major source of phosphorus, the other stimulant of cultural eutrophication.

There is no standard remedy for ailing lakes. Diversion is no cure-all. It has disadvantages in addition to its expense and capacity for creating or aggravating water problems elsewhere. Diversion can negate itself, particularly in water-scarce regions where sources of clean water that might have flushed lakes clean have been channeled instead into diversion systems to carry away sewage effluent. Thus, the advantage of diverting effluent is sometimes offset by the disadvantage of diverting potentially cleansing inflows. Engineers, aware of this drawback, are exploring the possibility of using "infertile" water from the Columbia River to rinse Moses Lake in central Washington.

Harvesting weeds and fish to remove nutrients from fertilized lakes promises to supplement diversion projects very effectively. Seining crews, for example, hauled 40 pounds of carp per acre from Madison's lake Mendota. Since the nitrogen and phosphorus content of fish flesh is 2.5 and 0.2 per cent, respectively, 500 pounds of carp harvested removed 12.5 pounds of nitrogen and a pound of phosphorus from the lakes. The 1966 carp catch yielded 16,000 pounds of nitrogen and 800 pounds of phosphorus. The nutrient "yield" of this method could be easily expanded in most eutrophic lakes with more intensive harvesting of carp and other rough fish.

A program to perfect weed-harvesting techniques and to evaluate their effectiveness was begun last summer on Lake Sallie, a densely populated lake in the vacation area of Detroit Lakes, Minn. By "cropping" one-third of the lake three times to a depth of five feet below the surface, the researchers expect to remove substantially more nutrients in one summer than enter the lake in a year. The outlook for harvesting algae on a large scale, however, isn't nearly as bright. No economical equipment has been developed to skim algae off lakes and impoundments.

While inventors are working on a variety of devices that "suck up algae like vacuum cleaners" or "strain algal scum from lakes," members of the National Eutrophication Research Program are hunting for a herbivore, perhaps some water flea or fish, with a huge appetite for algae. The search may be extended abroad if no algae-craving counterparts can be found to the weed-eating snails used in the southeastern United States. Precautions, however, must be taken to screen out "exotic" species that could become nuisances. Also, of two viruses known to cause diseases among blue-green algae, one is being tested, and more virulent species are being sought.

The best that can be said for spraying chemical poisons on lakes in the grip of algae and weeds is that it is usually a futile undertaking. Treating a lake with copper sulfate or other toxic chemicals is no more effective than taking aspirin for a brain tumor. It offers only temporary relief, masking the symptoms of cultural eutrophication. In the longer run it makes a lake sicker. Poisoning algae and weeds simply accelerates the natural process of growth, death, and decay, thereby freeing nutrients for another cycle of plant production. This has been borne out by studies of blue-green algae in Oregon's Upper Klamath Lake, which first showed signs of eutrophication eighty years ago. In the peak growth month of August, Klamath algae contain within their cells three times as much nitrogen as is dissolved in the lake. Killing an algae therefore would release this nitrogen for further growth.

Chemical poisons should be used only as a last resort, for once dumped into a body of water, they cannot be confined to one locality. They dissolve and spread far beyond the area treated. Dispersed eventually by wave action throughout a lake, they adversely alter the fragile fabric of aquatic communities of many species. Too little is known about the sublethal effects of such poisons to risk their use.

We can take several steps—all more effective than chemical poisoning—to bolster the two-pronged attack of diverting effluent and removing nutrients. By separating sewer systems for storm runoff and sewage, communities can forestall the lake fertilization that occurs when nutrient-rich combined sewers overflow after heavy rains. Damaging septic tank seepage can be curbed by passing legislation that would require tanks be installed at a "safe" distance from bodies of water. Wisconsin legislators, for instance, recently decided that lakes and rivers would be safe from fertilization only if septic tanks are set back as far as 1,000 feet and 300 feet, respectively. They also set more stringent specifications for septic tanks themselves. The specifications vary with the porosity of the soil.

Livestock growers could also fight cultural eutrophication by adopting the European practice of fluidizing and storing manure in vats, from winter freeze-up until the spring thaw, and then spraying it on their fields when the soil can better absorb it. The bulk of the manure now spread on frozen croplands runs off with spring meltwater to fertilize lakes and streams. This is significant considering the fact that the amount of manure, produced each year in the Midwest alone, is equivalent to human sewage of a population of 350 million. The chief drawback of the European method is its expense. Federal funds for water and soil conservation, however, could possibly be used to help farmers pay for costly storage tanks and spraying equipment.

We still have much to learn about the complexities of the eutrophication process. There is an urgent need for more research on such factors as a lake's depth, size, and configuration, its drainage basin and sources of nutrients, and its aquatic communities. But such research is complicated because biological and ecological factors vary with every lake, pond, and reservoir. Systems analysis with computers promises to be the research tool that will enable us to supply precise data on the potentialities of eutrophication.

Improved research, however, is only part of the solution to this dilemma. Unfortunately there isn't time to raise an enlightened generation to cope with eutrophication. Although we don't have all the answers, we must heed the warnings given by the waterways around us that have fallen into an advanced state of decay. If we delay, the price may be too dear to pay.

14.

Silent Spring

Rachel Carson

Rachel Carson is one of those rare biologists who combines good science with a superb literary style.

Her first book, *Under The Sea Wind,* was published in 1941. Ten years later, her National Book Award winner, *The Sea Around Us,* came into print. It was a best seller for eighty-six weeks and was eventually translated into thirty languages.

In all her work, Rachel Carson's basic interest and love has been the relation of life to its environment. From 1958 she had collected data from scientists all over the world about the dangerous effects of deadly poisons which we have unleashed upon our mother earth and all her children. The result was her famous *Silent Spring.* It is an elegant protest in behalf of the unity of all nature, a protest in behalf of life.

Elixirs of Death

For the first time in the history of the world, every human being is now subjected to contact with dangerous chemicals, from the moment of conception until death. In the less than two decades of their use, the synthetic pesticides have been so thoroughly distributed throughout the animate and inanimate world that they occur virtually everywhere. They have been recovered from most of the major river systems and even from streams of groundwater flowing unseen through the earth. Residues of these chemicals linger in soil to which they may have been applied a dozen years before. They have entered and lodged in the bodies of fish, birds, reptiles, and domestic and wild animals so universally that scientists carrying on animal experiments find it almost impossible to locate subjects free from such contamination. They have been found in fish in remote mountain lakes, in earthworms burrowing in soil, in the eggs of birds—and in man himself. For these chemicals are now stored in the bodies of the vast majority of human beings, regardless of age. They occur in the mother's milk, and probably in the tissues of the unborn child.

All this has come about because of the sudden rise and prodigious growth of an industry for the production of man-made or synthetic chemicals with insecticidal properties. This industry is a child of the Second World War. In the course of developing agents of chemical warfare, some of the chemicals created in the laboratory were found to be lethal to insects. The discovery did not come by chance: insects were widely used to test chemicals as agents of death for man.

The result has been a seemingly endless stream of synthetic insecticides. In being man-made—by ingenious laboratory manipulation of the molecules, substituting atoms, altering their arrangement—they differ sharply from the simpler inorganic insecticides of prewar days. These were derived from naturally occurring minerals and plant products—compounds of arsenic, copper, lead, manganese, zinc, and other minerals, pyrethrum from the dried flowers of chrysanthemums, nicotine sulphate from some of the relatives of tobacco, and rotenone from leguminous plants of the East Indies.

What sets the new synthetic insecticides apart is their enormous biological potency. They have immense power not merely to poison but to enter into the most vital processes of the body and change them in sinister and often deadly ways. Thus, as we shall see, they destroy the very enzymes whose function is to protect the body from harm, they block the oxidation processes from which the body receives its energy, they prevent the normal functioning of various organs, and they may initiate in certain cells the slow and irreversible change that leads to malignancy.

Yet new and more deadly chemicals are added to the list each year and new uses are devised so that contact with these materials has become practically worldwide. The production of synthetic pesticides in the United States soared from 124,259,000 pounds in 1947 to 637,666,000 pounds in 1960—more than a fivefold increase. The wholesale value of these products was well over a quarter of a billion dollars. But in the plans and hopes of the industry this enormous production is only a beginning.

A Who's Who of pesticides is therefore of concern to us all. If we are going to live so intimately with these chemicals—eating and drinking them, taking them into the very marrow of our bones—we had better know something about their nature and their power.

Although the Second World War marked a turning away from inorganic chemicals as pesticides into the wonder world of the carbon molecule, a few of the old materials persist. Chief among these is arsenic, which is still the basic ingredient in a variety of weed and insect killers. Arsenic is a highly toxic mineral occurring widely in association with the ores of various metals, and in very small amounts in volcanoes, in the sea, and in spring water. Its relations to man are varied and historic. Since many of its compounds are tasteless, it has been a favorite agent of homicide from long before the time of the Borgias to the present. Arsenic was the first recognized elementary carcinogen (or cancer-causing substance), identified in chimney soot and linked to cancer nearly two centuries ago by an English physician. Epidemics of chronic arsenical poisoning involving whole populations over long periods are on record. Arsenic-contaminated environments have also caused sickness and death among horses, cows, goats, pigs, deer, fishes, and bees; despite this record arsenical sprays and dusts are widely used. In the arsenic-sprayed cotton country of southern United States beekeeping as an industry has nearly died out. Farmers using arsenic dusts over long periods have been afflicted with chronic arsenic poisoning; livestock have been poisoned by crop sprays or weed killers containing arsenic. Drifting arsenic dusts from blueberry lands have spread over neighboring farms, contaminating streams, fatally poisoning bees and cows, and causing human illness. "It is scarcely possible .. to handle arsenicals with more utter disregard of the general health than that which has been practiced in our country in recent years," said Dr. W. C. Hueper, of the National Cancer Institute, an authority on environmental cancer. "Anyone who has watched the dusters and sprayers of arsenical insecticides at work must have been impressed by the almost supreme carelessness with which the poisonous substances are dispensed."

Modern insecticides are still more deadly. The vast majority fall into one of two large groups of chemicals. One, represented by DDT, is known as the "chlorinated hydrocarbons." The other group consists of the organic phosphorus insecticides, and is represented by the reasonably familiar malathion and parathion. All have one thing in common. As mentioned above, they are built on the basis of carbon atoms, which are also the indispensable building blocks of the living world, and thus classed as "organic." To understand them, we must see of what they are made, and how, although linked with the basic chemistry of all life, they lend themselves to the modifications which make them agents of death.

The basic element, carbon, is one whose atoms have an almost infinite capacity for uniting with each other in chains and rings and various other configurations, and for becoming linked with atoms of other substances. Indeed, the incredible diversity of living creatures from bacteria to the great blue whale is largely due to this capacity of carbon. The complex protein molecule has the carbon atom as its basis, as have molecules of fat, carbohydrates, enzymes, and vitamins. So, too, have enormous numbers of nonliving things, for carbon is not necessarily a symbol of life.

Some organic compounds are simply combinations of carbon and hydrogen. The simplest of these is methane, or marsh gas, formed in nature by the bacterial decomposition of organic matter under water. Mixed with air in proper proportions, methane becomes the dreaded "fire damp" of coal mines. Its structure is beautifully simple, consisting of one carbon atom to which four hydrogen atoms have become attached:

H H
C
H H

Chemists have discovered that it is possible to detach one or all of the hydrogen atoms and substitute other elements. For example, by substituting one atom of chlorine for one of hydrogen we produce methyl chloride:

H Cl
C
H H

Take away three hydrogen atoms and substitute chlorine and we have the anesthetic chloroform:

H Cl
C
Cl Cl

Substitute chlorine atoms for all of the hydrogen atoms and the result is carbon tetrachloride, the familiar cleaning fluid:

Cl Cl
C
Cl Cl

In the simplest possible terms, these changes rung upon the basic molecule of methane illustrate what a chlorinated hydrocarbon is. But this illustration gives little hint of the true complexity of the chemical world of the hydrocarbons, or of the manipulations by which the organic chemist creates his infinitely varied materials. For instead of the simple methane molecule with its single carbon atom, he may work with hydrocarbon molecules consisting of many carbon atoms, arranged in rings or chains, with side chains or branches, holding to themselves with chemical bonds not merely simple atoms of hydrogen or chlorine but also a wide variety of chemical groups. By seemingly slight changes the whole character of the substance is changed; for example, not only what is attached but the place of attachment to the carbon atom is highly important. Such ingenious manipulations have produced a battery of poisons of truly extraordinary power.

DDT (short for dichloro-diphenyl-tri-chloro-ethane) was first synthesized by a German chemist in 1874, but its properties as an insecticide were not discovered until 1939. Almost immediately DDT was hailed as a means of stamping out insect-borne disease and winning the farmers' war against crop destroyers overnight. The discoverer, Paul Müller of Switzerland, won the Nobel Prize.

DDT is now so universally used that in most minds the product takes on the harmless aspect of the familiar. Perhaps the myth of the harmlessness of DDT rests on the fact that one of its first uses was the wartime dusting of many thousands of soldiers, refugees, and prisoners, to combat lice. It is widely believed that since so many people came into extremely intimate contact with DDT and suffered no immediate ill effects the chemical must certainly be innocent of harm. This understandable misconception arises from the fact that—unlike other chlorinated hydrocarbons—ddt *in powder form* is not readily absorbed through the skin. Dissolved in oil, as it usually is, DDT is definitely toxic. If swallowed, it is absorbed slowly through the digestive tract; it may also be absorbed through the lungs. Once it has entered the body it is stored largely in organs rich in fatty substances (because DDT itself is fat-soluble) such as the adrenals, testes, or thyroid. Relatively large amounts are deposited in the liver, kidneys, and the fat of the large, protective mesenteries that enfold the intestines.

This storage of DDT begins with the smallest conceivable intake of the chemical (which is present as residues on most foodstuffs) and continues until quite high levels are reached. The fatty storage depots act as biological magnifiers, so that an intake of as little as 1/10 of 1 part per million in the diet results in storage of about 10 to 15 parts per million, an increase of one hundredfold or more. These terms of reference, so commonplace to the chemist or the pharmacologist, are unfamiliar to most of us. One part in a million sounds like a very small amount—and so it is. But such substances are so potent that a minute quantity can bring about vast changes in the body. In animal experiments, 3 parts per million has been found to inhibit an essential enzyme in heart muscle; only 5 parts per million has brought about necrosis or disintegration of liver cells; only 2.5 parts per million of the closely related chemicals dieldrin and chlordane did the same.

This is really not surprising. In the normal chemistry of the human body there is just such a disparity between cause and effect. For example, a quantity of iodine as small as two ten-thousandths of a gram spells the difference between health and disease. Because these small amounts of pesticides are cumulatively stored and only slowly excreted, the threat of chronic poisoning and degenerative changes of the liver and other organs is very real.

Scientists do not agree upon how much DDT can be stored in the human body. Dr. Arnold Lehman, who is the chief pharmacologist of the Food and Drug Administration, says there is neither a floor below which DDT is not absorbed nor a ceiling beyond which absorption and storage ceases. On the other hand, Dr. Wayland Hayes of the United States Public Health Service contends that in every individual a point of equilibrium is reached, and that DDT in excess of this amount is excreted. For practical purposes it is not particularly important which of these men is right. Storage in human beings has been well investigated, and we know that the average person is storing potentially harmful amounts. According to various studies, individuals with no known exposure (except the inevitable dietary one) store an average of 5.3

parts per million to 7.4 parts per million; agricultural workers 17.1 parts per million; and workers in insecticide plants as high as 648 parts per million! So the range of proven storage is quite wide and, what is even more to the point, the minimum figures are above the level at which damage to the liver and other organs or tissues may begin.

One of the most sinister features of DDT and related chemicals is the way they are passed on from one organism to another through all the links of the food chains. For example, fields of alfalfa are dusted with DDT; meal is later prepared from the alfalfa and fed to hens; the hens lay eggs which contain DDT. Or the hay, containing residues of 7 to 8 parts per million, may be fed to cows. The DDT will turn up in the milk in the amount of about 3 parts per million, but in butter made from this milk the concentration may run to 65 parts per million. Through such a process of transfer, what started out as a very small amount of DDT may end as a heavy concentration. Farmers nowadays find it difficult to obtain uncontaminated fodder for their milk cows, though the Food and Drug Administration forbids the presence of insecticide residues in milk shipped in interstate commerce.

The poison may also be passed on from mother to offspring. Insecticide residues have been recovered from human milk in samples tested by Food and Drug Administration scientists. This means that the breast-fed human infant is receiving small but regular additions to the load of toxic chemicals building up in his body. It is by no means his first exposure, however: there is good reason to believe this begins while he is still in the womb. In experimental animals the chlorinated hydrocarbon insecticides freely cross the barrier of the placenta, the traditional protective shield between the embryo and harmful substances in the mother's body. While the quantities so received by human infants would normally be small, they are not unimportant because children are more susceptible to poisoning than adults. This situation also means that today the average individual almost certainly starts life with the first deposit of the growing load of chemicals his body will be required to carry thenceforth.

All these facts—storage at even low levels, subsequent accumulation, and occurrence of liver damage at levels that may easily occur in normal diets, caused Food and Drug Administration scientists to declare as early as 1950 that it is "extremely likely the potential hazard of DDT has been underestimated." There has been no such parallel situation in medical history. No one yet knows what the ultimate consequences may be.

Chlordane, another chlorinated hydrocarbon, has all these unpleasant attributes of DDT plus a few that are peculiarly its own. Its residues are long persistent in soil, on foodstuffs, or on surfaces to which it may be applied. Chlordane makes use of all available portals to enter the body. It may be absorbed through the skin, may be breathed in as a spray or dust, and of course is absorbed from the digestive tract if residues are swallowed. Like all other chlorinated hydrocarbons, its deposits build up in the body in cumulative fashion. A diet containing such a small amount of chlordane as 2.5 parts per million may eventually lead to storage of 75 parts per million in the fat of experimental animals.

So experienced a pharmacologist as Dr. Lehman has described chlordane in 1950 as "one of the most toxic of insecticides—anyone handling it could be poisoned." Judging by the carefree liberality with which dusts for lawn treatments by suburbanites are laced with chlordane, this warning has not been taken to heart. The fact that the suburbanite is not instantly stricken has little meaning, for the toxins may sleep long in his body, to become manifest months or years later in an obscure disorder almost impossible to trace to its origins. On the other hand, death may strike quickly. One victim who accidentally spilled a 25 per cent industrial solution on the skin developed symptoms of poisoning within 40 minutes and died before medical help could be obtained. No reliance can be placed on receiving advance warning which might allow treatment to be had in time.

Heptachlor, one of the constituents of chlordane, is marketed as a separate formulation. It has a particularly high capacity for storage in fat. If the diet contains as little as 1/10 of 1 part per million there will be measurable amounts of heptachlor in the body. It also has the curious ability to undergo change into a chemically distinct substance known as heptachlor epoxide. It does this in soil and in the tissues of both plants and animals. Tests on birds indicate that the epoxide that results from this change is more toxic than the original chemical, which in turn is four times as toxic as chlordane.

As long ago as the mid-1930's a special group of hydrocarbons, the chlorinated naphthalenes, was found to cause hepatitis, and also a rare and almost invariably fatal liver disease in persons subjected to occupational exposure. They have led to illness and death of workers in electrical industries; and more recently, in agriculture, they have been considered a cause of a mysterious and usually fatal disease of cattle. In view of these antecedents, it is not surprising that three of the insecticides that are related to this group are among the most violently poisonous of all the hydrocarbons. These are dieldrin, aldrin, and endrin.

Dieldrin, named for a German chemist, Diels, is about 5 times as toxic as DDT when swallowed but 40 times as toxic when absorbed through the skin in solution. It is notorious for striking quickly and with terrible effect at the nervous system, sending the victims into convulsions. Persons thus poisoned recover so slowly as to indicate chronic effects. As with other chlorinated hydrocarbons, these long-term effects include severe damage to the liver. The long duration of its residues and the effective insecticidal action make dieldrin one of the most used insecticides today, despite the appalling destruction of wildlife that has followed its use. As tested on quail and pheasants, it has proved to be about 40 to 50 times as toxic as DDT.

There are vast gaps in our knowledge of how dieldrin is stored or distributed in the body, or excreted, for the chemists' ingenuity in devising insecticides has long ago outrun biological knowledge of the way these poisons affect the living organism. However, there is every indication of long storage in the human body, where deposits may lie dormant like a slumbering volcano, only to flare up in periods of physiological stress when the body draws upon its fat reserves. Much of what we do know has been learned through hard experience in the antimalarial campaigns carried out by the World Health Organization. As soon as dieldrin was substituted for DDT in malaria-control work (because the malaria mosquitoes had become resistant to DDT), cases of poisoning among the spraymen began to occur. The seizures were severe—from half to all (varying in the different programs) of the men affected went into convulsions and several died. Some had convulsions as long as *four months* after the last exposure.

Aldrin is a somewhat mysterious substance, for although it exists as a separate entity it bears the relation of alter ego to dieldrin. When carrots are taken from a bed treated with aldrin they are found to contain residues of dieldrin. This change occurs in living tissues and also in soil. Such alchemistic transformations have led to many erroneous reports, for if a chemist, knowing aldrin has been applied, tests for it he will be deceived into thinking all residues have been dissipated. The residues are there, but they are dieldrin and this requires a different test.

Like dieldrin, aldrin is extremely toxic. It produces degenerative changes in the liver and kidneys. A quantity the size of an aspirin tablet is enough to kill more than 400 quail. Many cases of human poisonings are on record, most of them in connection with industrial handling.

Aldrin, like most of this group of insecticides, projects a menacing shadow into the future, the shadow of sterility. Pheasants fed quantities too small to kill them nevertheless laid few eggs, and the chicks that hatched soon died. The effect is not confined to birds. Rats exposed to aldrin had fewer pregnancies and their young were sickly and short-lived. Puppies born of treated mothers died within three days. By one means or another, the new generations suffer for the posoning of their parents. No one knows whether the same effect will be seen in human beings, yet this chemical has been sprayed from airplanes over suburban areas and farmlands.

Endrin is the most toxic of all the chlorinated hydrocarbons. Although chemically rather closely related to dieldrin, a little twist in its molecular structure makes it 5 times as poisonous. It makes the progenitor of all this group of insecticides, DDT, seem by comparison almost harmless. It is 15 times as poisonous as DDT to mammals, 30 times as poisonous to fish, and about 300 times as poisonous to some birds.

In the decade of its use, endrin has killed enormous numbers of fish, has fatally poisoned cattle that have wandered into sprayed orchards, has poisoned wells, and has drawn a sharp warning from at least one state health department that its careless use is endangering human lives.

In one of the most tragic cases of endrin posoning there was no apparent carelessness; efforts had been made to take precautions apparently considered adequate. A year-old child had been taken by his American parents to live in Venezuela. There were cockroaches in the house to which they moved, and after a few days a spray containing endrin was used. The baby and the small family dog were taken out of the house before the spraying was done about nine o'clock one morning. After the spraying the floors were washed. The baby and dog were returned to the house in midafternoon. An hour or so later the dog vomited, went into convulsions, and died. At 10 P.M. on the evening of the same day the baby also vomited, went into convulsions, and lost consciousness. After that fateful contact with endrin, this normal, healthy child became little more than a vegetable—unable to see or hear, subject to frequent muscular spasms, apparently completely cut off from contact with his surroundings. Several months of treatment in a New York hospital failed to change his condition or bring hope of change. "It is extremely doubtful," reported the attending physicians, "that any useful degree of recovery will occur."

The second major group of insecticides, the alkyl or organic phosphates, are among the most poisonous chemicals in the world. The chief and most obvious hazard attending their use is that of acute poisoning of people applying the sprays or accidentally coming in contact with drifting spray, with vegetation coated by it, or with a discarded container. In Florida, two children found an empty bag and used it to repair a swing. Shortly thereafter both of them died and three of their playmates became ill. The bag had once contained an insecticide called parathion, one of the organic phosphates; tests established death by parathion poisoning. On another occasion two small boys in Wisconsin, cousins, died on the same night. One had been playing in his yard when spray drifted in from an adjoining field where his father was spraying potatoes with parathion; the other had run playfully into the barn after his father and had put his hand on the nozzle of the spray equipment.

The origin of these insecticides has a certain ironic significance. Although some of the chemicals themselves—organic esters of phosphoric acid—had been known for many years, their insecticidal properties remained to be discovered by a German chemist, Gerhard Schrader, in the late 1930's. Almost immediately the German government recognized the value of these same chemicals as new and devastating weapons in man's war against his own kind, and the work on them was declared secret. Some became the deadly nerve gases. Others, of closely allied structure, became insecticides.

The organic phosphorus insecticides act on the living organism in a peculiar way. They have the ability to destroy enzymes—enzymes that perform necessary functions in the body. Their target is the nervous system, whether the victim is an insect or a warm-blooded animal. Under normal conditions, an impulse passes from nerve to nerve with the aid of a "chemical transmitter" called acetylcholine, a substance that performs an essential function and then disappears. Indeed, its existence is so ephemeral that medical researchers are unable, without special procedures, to sample it before the body has destroyed it. This transient nature of the transmitting chemical is necessary to the normal functioning of the body. If the acetylcholine is not destroyed as soon as a nerve impulse has passed, impules continue to flash acorss the bridge from nerve to nerve, as the chemical exerts its effects in an ever more intensified manner. The movements of the whole body become uncoordinated: tremors, muscular spasms, convulsions, and death quickly result.

This contingency has been provided for by the body. A protective enzyme called cholinesterase is at hand to destroy the transmitting chemical once it is no longer needed. By this means a precise balance is struck and the body never builds up a dangerous amount of acetylcholine. But on contact with the organic phosphorus insecticides, the protective enzyme is destroyed, and as the quantity of the enzyme is reduced that of the transmitting chemical builds up. In this effect, the organic phosphorus compounds resemble the alkaloid poison muscarine, found in a poisonous mushroom, the fly amanita.

Repeated exposures may lower the cholinesterase level until an individual reaches the brink of acute poisoning, a brink over which he may be pushed by a very small additional exposure. For this reason it is considered important to make periodic examinations of the blood of spray operators and others regularly exposed.

Parathion is one of the most widely used of the organic phosphates. It is also one of the most powerful and dangerous. Honeybees become "wildly agitated and bellicose" on contact with it, perform frantic cleaning movements, and are near death within half an hour. A chemist, thinking to learn by the most direct possible means the dose acutely toxic to human beings, swallowed a minute amount, equivalent to about .00424 ounce. Paralysis followed so instantaneously that he could not reach the antidotes he had prepared at hand, and so he died. Parathion is now said to be a favorite instrument of suicide in Finland. In recent years the State of California has reported an average of more than 200 cases of accidental parathion poisoning annually. In many parts of the world the fatality rate from parathion is startling: 100 fatal cases in India and 67 in Syria in 1958, and an average of 336 deaths per year in Japan.

Yet some 7,000,000 pounds of parathion are now applied to fields and orchards of the United States—by hand sprayers, motorized blowers and dusters, and by airplane. The amount used on California farms alone could, according to one medical authority, "provide a lethal dose for 5 to 10 times the whole world's population."

One of the few circumstances that save us from extincition by this means is the fact that parathion and other chemicals of this group are decomposed rather rapidly. Their residues on the crops to which they are applied are therefore relatively short-lived compared with the chlorinated hydrocarbons. However, they last long enough to create hazards and produce consequences that range from the merely serious to the fatal. In Riverside, California, eleven out of thirty men picking oranges became violently ill and all but one had to be hospitalized. Their symptoms were typical of parathion poisoning. The grove had been sprayed with parathion some two and a half weeks earlier; the residues that reduced them to retching, half-blind, semiconscious misery were sixteen to nineteen days old. And this is not by any means a record for persistence. Similar mishaps have occurred in groves sprayed a month earlier, and residues have been found in the peel of oranges six months after treatment with standard dosages.

The danger to all workers applying the organic phosphorus insecticides in fields, orchards, and vineyards, is so extreme that some states using these chemicals have established laboratories where physicians may obtain aid in diagnosis and treatment. Even the physicians themselves may be in some danger, unless they wear rubber gloves in handling the victims of poisoning. So may a laundress washing the clothing of such victims, which may have absorbed enough parathion to affect her.

Malathion, another of the organic phosphates, is almost as familiar to the public as DDT, being widely used by gardeners, in household insecticides, in mosquito spraying, and in such blanket attacks on insects the spraying of nearly a million acres of Florida communities for the Mediterranean fruit fly. It is considered the least toxic of this group of chemicals and many people assume they may use it freely and without fear of harm. Commercial advertising encourages this comfortable attitude.

The alleged "safety" of malathion rests on rather precarious ground, although—as often happens—this was not discovered until the chemical had been in use for several years. Malathion is "safe" only because the mammalian liver, an organ with extraordinary protective powers, renders it relatively harmless. The detoxification is accomplished by one of the enzymes of the liver. If, however, something destroys this enzyme or interferes with its action, the person exposed to malathion receives the full force of the poison.

Unfortunately for all of us, opportunities for this sort of thing to happen are legion. A few years ago a team of Food and Drug Administration scientists discovered that when malathion and certain other organic phosphates are administered simultaneously a massive poisoning results—up to 50 times as severe as would be predicted on the basis of adding together the toxicities of the two. In other words, 1/100 of the lethal dose of each compound may be fatal when the two are combined.

This discovery led to the testing of other combinations. It is now known that many pairs of organic phosphate insecticides are highly dangerous, the toxicity being stepped up or "potentiated" through the combined action. Potentiation seems to take place when one compound destroys the liver enzyme responsible for detoxifying the other. The two need not be given simultaneously. The hazard exists not only for the man who may spray this week with one insecticide and next week with another; it exists also for the consumer of sprayed products. The common salad bowl may easily present a combination of organic phosphate insecticides. Residues well within the legally permissible limits may interact.

The full scope of the dangerous interaction of chemicals is as yet little known, but disturbing findings now come regularly from scientific laboratories. Among these is the discovery that the toxicity of an organic phosphate can be increased by a second agent that is not necessarily an insecticide. For example, one of the plasticizing agents may act even more strongly than another insecticide to make malathion more dangerous. Again, this is because it inhibits the liver enzyme that normally would "draw the teeth" of the poisonous insecticide.

What of other chemicals in the normal human environment? What, in particular, of drugs? A bare beginning has been made on this subject, but already it is known that some organic phosphates (parathion and malathion) increase the toxicity of some drugs used as muscle relaxants, and that several others (again including malathion) markedly increase the sleeping time of barbiturates.

In Greek mythology the sorceress Medea, enraged at being supplanted by a rival for the affections of her husband Jason, presented the new bride with a robe possessing magic properties. The wearer of the robe immediately suffered a violent death. This death-by-indirection now finds its counterpart in what are known as "systemic insecticides." These are chemicals with extraordinary properties which are used to convert plants or animals into a sort of Medea's robe by making them actually poisonous. This is done with the purpose of killing insects that may come in contact with them, especially by sucking their juices or blood.

The world of systemic insecticides is a weird world, surpassing the imaginings of the brothers Grimm—perhaps most closely akin to the cartoon world of Charles Addams. It is a world where the enchanted forest of the fairy tales has become the poisonous forest in which an insect that chews a leaf or sucks the sap of a plant is doomed. It is a world where a flea bites a dog, and dies because the dog's blood has been made poisonous, where an insect may die from vapors emanating from a plant it has never touched, where a bee may carry poisonous nectar back to its hive and presently produce poisonous honey.

The entomologists' dream of the built-in insecticide was born when workers in the field of applied entomology realized they could take a hint from nature: they found that wheat growing in soil containing sodium selenate was immune to attack by aphids or spider mites. Selenium, a naturally occurring element found sparingly in rocks and soils of many parts of the world, thus became the first systemic insecticide.

What makes an insecticide a systemic is the ability to permeate all the tissues of a plant or animal and make them toxic. This quality is possessed by some chemicals of the chlorinated hydrocarbon group and by others of the organophosphorus group, all synthetically produced, as well as by certain naturally occurring substances. In practice, however, most systemics are drawn from the organophosphorus group because the problem of residues is somewhat less acute.

Systemics act in other devious ways. Applied to seeds, either by soaking or in a coating combined with carbon, they extend their effects into the following plant generation and produce seedlings poisonous to aphids and other sucking insects. Vegetables such as peas, beans, and sugar beets are sometimes thus protected. Cotton seeds coated with a systemic insecticide have been in use for some time in California, where 25 farm laborers planting cotton in the San Joaquin Valley in 1959 were seized with sudden illness, caused by handling the bags of treated seeds.

In England someone wondered what happened when bees made use of nectar from plants treated with systemics. This was investigated in areas treated with a chemical called schradan. Although the plants had been sprayed before the flowers were formed, the nectar later produced contained the poison. The result, as might have been predicted, was that the honey made by the bees also was contaminated with schradan.

Use of animal systemics has concentrated chiefly on control of the cattle grub, a damaging parasite of livestock. Extreme care must be used in order to create an insecticidal effect in the blood and tissues of the host without setting up a fatal poisoning. The balance is delicate and government veterinarians have found that repeated small doses can gradually deplete an animal's supply of the protective enzyme cholinesterase, so that without warning a minute additional dose will cause poisoning.

There are strong indications that fields closer to our daily lives are being opened up. You may now give your dog a pill which, it is claimed, will rid him of fleas by making his blood poisonous to them. The hazards discovered in treating cattle would presumably apply to the dog. As yet no one seems to have proposed a human systemic that would make us lethal to a mosquito. Perhaps this is the next step.

So far in this chapter we have been discussing the deadly chemicals that are being used in our war against the insects. What of our simultaneous war against the weeds?

The desire for a quick and easy method of killing unwanted plants has given rise to a large and growing array of chemicals that are known as herbicides, or, less formally, as weed killers. The story of how these chemicals are used and misused will be told in Chapter 6; the question that here concerns us is whether the weed killers are poisons and whether their use is contributing to the poisoning of the environment.

The legend that the herbicides are toxic only to plants and so pose no threat to animal life has been widely disseminated, but unfortunately it is not true. The plant killers include a large variety of chemicals that act on animal tissue as well as on vegetation. They vary greatly in their action on the organism. Some are general poisons, some are powerful stimulants of metabolism, causing a fatal rise in body temperature, some induce malignant tumors either alone or in partnership with other chemicals, some strike at the genetic material of the race by causing gene mutations. The herbicides, then, like the insecticides, include some very dangerous chemicals, and their careless use in the belief that they are "safe" can have disastrous results.

Despite the competition of a constant stream of new chemicals issuing from the laboratories, arsenic compounds are still liberally used, both as insecticides (as mentioned above) and as weed killers, where they usually take the chemical form of sodium arsenite. The history of their use is not reassuring. As roadside sprays, they have cost many a farmer his cow and killed uncounted numbers of wild creatures. As aquatic weed killers in lakes and reservoirs they have made public waters unsuitable for drinking or even for swimming. As a spray applied to potato fields to destroy the vines they have taken a toll of human and nonhuman life.

In England this latter practice developed about 1951 as a result of a shortage of sulfuric acid, formerly used to burn off the potato vines. The Ministry of Agriculture considered it necessary to give warning of the hazard of going into the arsenic-sprayed fields, but the warning was not understood by the cattle (nor, we must assume, by the wild animals and birds) and reports of cattle poisoned by the arsenic sprays came with monotonous regularity. When death came also to a farmer's wife through arsenic-contaminated water, one of the major English chemical companies (in 1959) stopped production of arsenical sprays and called in supplies already in the hands of dealers, and shortly thereafter the Ministry of Agriculture announced that because of high risks to people and cattle restrictions on the use of arsenites would be imposed. In 1961, the Australian government announced a similar ban. No such restrictions impede the use of these poisons in the United States, however.

Some of the "dinitro" compounds are also used as herbicides. They are rated as among the most dangerous materials of this type in use in the United States. Dinitrophenol is a strong metabolic stimulant. For this reason it was at one time used as a reducing drug, but the margin between the slimming dose and that required to poison or kill was slight—so slight that several patients died and many suffered permanent injury before use of the drug was finally halted.

A related chemical, pentachlorophenol, sometimes known as "penta," is used as a weed killer as well as an insecticide, often being sprayed along railroad tracks and in waste areas. Penta is extremely toxic to a wide variety of organisms from bacteria to man. Like the dinitros, it interferes, often fatally, with the body's source of energy, so that the affected organism almost literally burns itself up. Its fearful power is illustrated in a fatal accident recently reported by the California Department of Health. A tank truck driver was preparing a cotton defoliant by mixing diesel oil with pentachlorophenol. As he was drawing the concentrated chemical out of a drum, the spigot accidentally toppled back. He reached in with his bare hand to regain the spigot. Although he washed immediately, he became acutely ill and died the next day.

While the results of weed killers such as sodium arsenite or the phenols are grossly obvious, some other herbicides are more insidious in their effects. For example, the now famous cranberry-weed-killer aminotriazole, or amitrol, is rated as having relatively low toxicity. But in the long run its tendency to cause malignant tumors of the thyroid may be far more significant for wildlife and perhaps also for man.

Among the herbicides are some that are classified as "mutagens," or agents capable of modifying the genes, the materials of heredity. We are rightly appalled by the genetic effects of radiation; how then, can we be indifferent to the same effect in chemicals that we disseminate widely in our environment?

15.

The Economics of Wilderness

Garrett Hardin

Professor Garrett Hardin is currently Professor of Biology at the University of California, Santa Barbara. He is one of the most articulate and gifted writers in the field of science. For the past twenty-five years he has focused his energies on the social implications of biology. He has published widely in magazines and his books include *Biology: Its Human Implications*, *Nature and Man's Fate*, and *39 Steps to Biology.*

This magazine article, "The Economics of Wilderness," is a classic of its kind. It is based on a talk given by Garrett Hardin at the Sierra Club's Wilderness Conference, San Francisco, March 15, 1969.

To some it may seem anathema to mention wilderness and economics in the same breath. Certainly, in the past, some of the most dangerous enemies of wilderness have been men who spoke the economic lingo. Despite this historic war I think the brush of economics is a proper one for painting a picture of wilderness as a problem in human choice.

Economics may be defined as the study of choice necessitated by scarcity. There is something odd, and even improper, in speaking of the "economics of abundance" as Stuart Chase once did. With true abundance all economics ceases, except for the ultimately inescapable economics of time. Of the economics of time there is no general theory, and perhaps cannot be. But for the *things* of the world there is an economics, something that can be said.

Although there really is no such thing as an economics of abundance, the belief that there is, is one of the suppurating myths of our time. This belief had its origin partly in a genuine economic phenomenon, "the economy of scale." For complex artifacts in general the unit cost goes down as the scale of manufacture increases. In general, the more complex the artifact, the more striking the economy of scale: the cost per unit to build a million automobiles per year is far, far less than the cost per unit when only one is manufactured. Because artifacts are so pervasive in modern life, most of us unconsciously assume "the bigger the better," and "the more the cheaper." It takes a positive effort of imagination to realize that there are things the supply of which cannot be multiplied indefinitely. Natural resources in general, and wilderness in particular, fall in this group.

This is obvious enough to Sierra Club members. It should be obvious to everyone, but it is not. Not long ago, for example, discussing some propsoed improvements in a national park, the *Toronto Financial Post* said: "During 1968 and early 1969, campsites will be expanded and roads paved to enable the visitor to enjoy the wilderness atmosphere that was nearly inaccessible only a few years ago." This is an astonishing sentence, but I will bet that one would have to argue with the writer of it for quite a while before he could be made to see the paradox involved in speaking of building a road into the wilderness.

Wilderness cannot be multiplied, and it can be subdivided only a little. It is not increasing; we have to struggle to keep it from decreasing as population increases. Were we to divide up the wilderness among even a small fraction of the total population, there would be no real wilderness available to anyone. So what should we do?

The first thing to do is to see where we stand, to make a list of possibilities without (initially) making any judgment of their desirability. On the first level of analysis there are just three possibilities.

1. The wilderness can be opened to everyone. The end result of this is completely predictable: absolute destruction. Only a nation with a small population, perhaps no greater than one percent of our present population, a nation that does not have at its disposal our present means of transportation could maintain a wilderness that was open to all.

2. We can close the wilderness to everyone. In a limited sense, this action would preserve the wilderness. But it would be a wilderness like Bishop Berkeley's "tree in the quad" when no one is there: does wilderness really exist if no one experiences it? Such an action would save wilderness for the future, but it would do no one any good now.

3. We can allow only limited access to the wilderness. This is the only course of action that can be rationally defended. Only a small percentage of a large population can ever enjoy wilderness. By suitably defining our standards, and by studying the variables in the situation, we can (in principle) work out a theory for maximizing the enjoyment of wilderness under a system of limited access. Whatever our theory, we shall have to wrestle with the problem of choice, the problem of determining what small number among a vast population of people shall have the opportunity to enjoy this scarce good, wilderness. It is this problem of choice that I wish to explore here.

What I have to say applies not only to wilderness in the sense in which that term is understood by all good outdoorsmen, but also to all other kinds of outdoor recreational areas—to national parks, to ski areas, and the like. All of these can be destroyed by localized overpopulation. They differ in their "carrying capacity." The carrying capacity of a Coney Island (for those who like it, and there are such people) is very high; the carrying capacity of wilderness, in the sense defined by Howard Zahniser, is very low. In the Wilderness Bill of 1964 Zahniser's felicitous definition stands for all to admire:

"A wilderness, in contrast with those areas where man and his own works dominate the landscape is hereby recognized as an area where the earth and its community of life are untrammeled by man, where man himself is a visitor who does not remain."

The carrying capacity of Coney Island is, I suppose, something like 100 people per acre; the carrying capacity of a wilderness is perhaps one person per square mile. But whatever the carrying capacity, as population inexorably increases, each type of recreational area sooner or later comes up against the problem of allocation of this scarce resource among the more than sufficient number of claimants to it. It is at this point that the problem of limited access must be faced.

How shall we limit access? How shall we choose from among the too-abundant petitioners those few who shall be allowed in? Let's run over the various possibilities.

First: By the marketplace. We can auction off the natural resource, letting those who are richest among the sufficiently motivated buy. In our part of the world and in our time most of us unhesitatingly label this method of allotment "unfair." Perhaps it is. But don't forget that many an area of natural beauty available to us today has survived unspoiled precisely because it was preserved in an estate of the wealthy in past times. This method of allotment has at least the virtue that it preserves natural treasures until a better, or perhaps we should merely say a more acceptable, method of distribution can be devised. The privilege of wealth has in the past carried many of the beauties of nature through the first, destructive eras of nascent democracy to the more mature, later stages that were capable of appreciating and preserving them.

Second: By queues. Wilderness could be made available on a first-come, first-served basis, up to the extent of the carrying capacity. People would simply line up each day in a long queue and a few would be allowed in. It would be a fatiguing and wasteful system, but while it would be "fair," it might not be stable.

Third: By lottery. This would be eminently "fair," and it would not be terribly fatiguing or wasteful. In earlier days, the decision of a lottery was regarded as the choice of God. We cannot recapture this consoling belief (now that "God is dead"), but we are still inclined to accept the results of a lottery. Lotteries serve well for the allocation of hunting rights in some of our states where big game abounds.

Fourth: By merit. Whether one regards this as "unfair" or "fair" depends on the complexion of one's political beliefs. Whether it is fair or not, I will argue that it is the best system of allocation. Anyone who argues for a merit system of determining rights immediately raises an *argumentum ad hominem.* He immediately raises the suspicion that he is about to define merit in such a way as to include himself in the meritorious group.

The suspicion is justified, and because it it justified it must be met. To carry conviction, he who proposes standards must show that his argument is not self-serving. What I hereby propose as a criterion for admission to the wilderness is great physical vigor. I explicitly call your attention to this significant fact: I myself cannot pass the test I propose. I had polio at the age of four, and got around moderately well for more than 40 years, but now I require crutches. Until today, I have not traded on my infirmity. But today I must, for it is an essential part of my argument.

I am not fit for the wilderness I praise. I cannot pass the test I propose. I cannot enter the area I would restrict. Therefore I claim that I speak with objectivity. The standard I propose is not an example of special pleading in my own interest. I can speak loudly where abler men would have to be hesitant.

To restrict the wilderness to physically vigorous people is inherently sensible. What is the experience of wilderness? Surely it has two major components. The first is the experience of being there, of (in Thoreau's words) being refreshed "by the sight of inexhaustible vigor, vast and titanic features," of seeing "that nature is so rife with life that myriads can afford to be sacrificed and suffered to prey on one another; that tender organizations can be so serenely squashed out of existence like pulp. . . . "

The experience of being there is part of the experience of wilderness, but only a part. If we were dropped down from a line by helicopter into the middle of this experience we would miss an important part of the total experience, namely the experience of getting there. The exquisite sight, sound, and smell of wilderness is many times more powerful if it is earned through physical achievement, if it comes at the end of a long and fatiguing trip for which vigorous good health is a necessity.

Practically speaking, this means that no one should be able to enter a wilderness by mechanical means. He should have to walk many miles on his own two feet, carrying all his provisions with him. In some cases, entrance might be on horse or mule back, or in a canoe, or by snowshoes; but there should be no automobiles, no campers, no motorcycles, no tote-goats, no outboard motors, no airplanes. Just unmechanized man and nature—this is a necessary ingredient of the prescription for the wilderness experience.

That mechanical aids threaten wilderness is already recognized by managers of our wildernesses. Emergency roads, it is said, should be used sparingly. I submit that this cautious policy is not cautious enough. I submit that there should be *no* emergency roads, that the people who go into the wilderness should go in without radio transmitters, that they should know for certain that if an emergency arises they can get no help from the outside. If injured, they must either somehow struggle to the outside under their own power or (if lucky) catch the attention of another rare wanderer in the wilderness and get him to help. For people who are physically prepared for it, the wilderness is not terribly dangerous—but such danger as there is, is a precious part of the total experience. The knowledge that one is really on one's own is a powerful tonic. It would be cruelly sentimental to take this away from the wilderness adventurer.

There is not even a public interest in making the wilderness safe. Making great and spectacular efforts to save the life of an individual makes sense only when there is a shortage of people. I have not lately heard that there is a shortage of people.

There is, however, a public interest in making the wilderness as difficult and dangerous as it legitimately can be. There is, I think a well-founded suspicion that our life has become, if anything, too safe for the best psychological health, particularly among the young. The ever greater extension of the boundaries of legal liability has produced a controlled and fenced-in environment in which it is almost impossible to hurt oneself—unless one tries. The behavior of the young clearly indicates that they really try. Drag races, road races, "rumble," student sit-ins, marches, and tauntings of the police—all these activities look like the behavior of people looking for danger. I do not wish to deny that some of the activities may arise from other motivations, also, e.g., idealistic political beliefs. I am only saying that it looks like deliberate seeking of danger is part of the motivation of our obstreperous young. I think it is an important part. I think we would do well to tear down some of the fences that now deprive people of the possibility of danger. A wilderness without rescue services would contribute to the stability of society.

There is a second way in which the interest of society is furthered by a rigorous wilderness. From time to time a president of the United States endeavors to improve the physical condition of the average citizen by resorting to a rhetorical bombardment. The verbal ammunition consists principally of the words "responsibility," "duty," and "patriotism." These rhetorical duds no longer move the young. The negative motivation of shame is, in general, not as effective as the positive motivation of prestige. A wilderness that can be entered only by a few of the most physically fit of the population will act as an incentive to myriads more to improve their physical condition. The motivation will be more effective if we have (as I think we should) a graded series of wilderness and park areas. Areas in which the carrying capacity is reckoned at one person per thousand acres should be the most difficult to enter; those with a capacity of one per hundred acres should be easier, those with one per ten, still easier, and so on. Yosemite Valley should, I suggest, be assigned a carrying capacity of about one per acre which might mean that it could be opened to anyone who could walk ten miles. At first, of course, the ten-mile walkers would be a very small class, but once the prestige factor took effect more and more people would be willing to walk such a distance. Then the standard should be made more rigorous.

I am sure other details of such a system would eventually have to be faced and worked out. It might be necessary to combine it with a lottery. Or some independent, easily administered test of physical fitness might be instituted. These are details, and in principle can be solved, so I will not spend time on them. But whatever the details, it is clear that many of our present national park and national forests and other recreation areas should be forever closed to people on crutches, to small children, to fat people, to people with heart condition, and to old people in the usual state of physical disrepair. On the basis of their lack of merit, such people (and remember, I am a member of this deprived group) should give up all claim of right to the wilderness experience.

The poet Goethe once said, "We must earn again for ourselves what we have inherited," recognizing that only those things that are earned can be precious. To be precious the heritage of wilderness must be open only to those who can earn it again for themselves. The rest, since they cannot gain the genuine treasure by their own efforts, must relinquish the shadow of it.

We need not be so righteous as to deny the excluded ones all experience of the out-of-doors. There is no reason in the world why we cannot expand our present practice of setting up small outdoor areas where we permit a high density of people to get a tiny whiff of nature. Camping cheek by jowl with thousands of others in an outdoor slum does not appeal to me personally—I have not visited Yosemite Valley in thirty years—but there are people who simply love this slummy togetherness, a fact that Sierra Clubbers sometimes forget or find hard to believe. By all means, let us create some al fresco slums for the people, but not in the likes of Yosemite Valley, which is too good for this purpose. But there will be little loss if some of the less attrctive forest areas are turned into outdoor slums to relieve the pressure on the really good areas. We must have lakes that fairly pullulate with water skiers in order that we may be able to set aside other lakes for quiet canoeing. We must have easily reached beaches that fairly writhe with oily bodies and vibrate to a steady cacophony of transistor radios, in order to keep up other beaches, difficult of access, on which we can forbid all noise makers.

The idea of wilderness is a difficult one, but it is precisely because it is difficult that clarifying it is valuable. In discovering how to justify a restricted good to a nation of 200 million people that is still growing, we find a formula that extends beyond wilderness to a whole spectrum of recreational activities in the national commons. The solution of the difficult case erects a framework into which other cases can be easily fitted.

All Mankind Has a Stake

Robert Cahn

Robert Cahn is a gifted writer for *Christian Science Monitor*. His series of fifteen articles entitled "Will Success Spoil the National Parks?" has captured several awards and elicited so many requests for reprints that it was put into book form.

In this excerpt, "All Mankind Has a Stake," Mr. Cahn states with forthrightness the need to preserve areas of natural beauty for all of humanity.

"It is truly magnificent," a tall, gracious visitor from Bombay commented one day last fall as he looked into the Grand Canyon.

"But . . . "

The visitor paused as he gazed from the south rim into the yawning, mile-deep chasm carved by nature's forces millions of years ago. His deep-set eyes absorbed the shadings of red, yellow, and blue as the late afternoon clouds sent shadows running across the buttes and spires inside the canyon.

"When your Congress provided that natural areas like this should be maintained unimpaired for future generations of Americans, that was just the first step," he continued. "The Grand Canyon is more than American—it should be preserved for all the world."

Zafar Futehally is honorary secretary of the Bombay Natural History Society and a leader in the movement to emphasize international values of parks of all nations. Mr. Futehally had come to the United States with 34 other representatives from 25 countries. Grand Canyon National Park was the final stop in a four-week course in administration of national parks and conservation areas.

Old Concept Stretched

The concept of each country's international responsibility for preserving its unique natural wonders adds a new dimension to the conservation concepts of many Americans. The time-honored United States viewpoint was perhaps best set forth by the nation's foremost conservation president, Theodore Roosevelt.

"Leave it as it is," said President Roosevelt when he first viewed the Grand Canyon in 1903. "You cannot improve on it. The ages have been at work on it, and man can only mar it. What you can do is to keep it for your children, your children's children, and for all who come after you as one of the great sights which every American . . . can see."

Mr. Futehally and other internationally minded conservationists recognize the effects of the transportation and communications revolutions of the 20th century. The world has shrunk. Millions of foreigners now have heard about and seen pictures and even television views of the Grand Canyon or the Florida Everglades National Parks.

Tourists Crisscross

Each year, thousands come to the United States to visit these and other scenic spots. So do thousands of Americans travel to such outstanding areas as Iguassú Falls (Argentina-Brazil), or to the volcanic cone of Japan's Mt. Fuji, or to the spectacular wildlife display of Kenya's Amboseli-Masai game reserve in the shadow of Mt. Kilimanjaro.

Areas of this caliber, unique in the world, should thus be given priority for preservation, says Mr. Futehally. But in the press of competing national demands, the high-sounding principles of conservation do not always win out over the pressures for industrial, agricultural, or commercial and urban development.

Mr. Futehally was too polite to mention specifics. But it is no secret that two of America's greatest natural attractions, the Grand Canyon and Everglades National Park, have in the past few years narrowly escaped extensive man-caused interference. And they may be threatened again.

Wildlife Declines

In Everglades National Park, a series of flood-control gates and canals constructed by the United States Army Corps of Engineers interfered seriously with the normal flow of water into the Everglades. The adverse effects on the park were intensified during a period of severe drought. Water that normally would have gone into the Everglades park even in the drought years went instead to southern Florida cities and farms, or was discharged directly into the ocean.

Much of the park's wildlife suffered. Reproduction of wading birds declined drastically, the total dropping from 1.5 million in the 1930's to less than 50,000 today. It is also estimated that the number of alligators has declined 95 percent since the 1920's. This reflects the delicate relationship between the amount of water and the abundance of plants and animals on which the birds and alligators depend. (Part of the alligator loss has been due to poaching.)

After complaints from conservationists and the National Park Service, the Corps of Engineers and the Central and South Florida Flood Control Districts agreed to give the Everglades park additional water. The situation has been critical during several recent drought periods. But a written agreement to provide the park with 325,000 acre-feet of water a year, regardless of the increasing domestic demands in Florida, is in process of being worked out now.

Canyon Dam Defeated

Two years ago, Grand Canyon National Park became caught in the cross fire of legislation which would have permitted a large hydroelectric dam on the Colorado River just below the park. Areas in the canyon's depths, set aside for their scenic grandeur, would have been flooded as the river backed up behind the dam.

Conservation groups battled the supporters of the dam and forced Congress to listen. At present, the advocates of the dam have lost out, although they have not given up the fight.

For Mr. Futehally and the other participants in the third international course, the pressures of overdevelopment and overcrowding of park areas are not yet imminent dangers in most of their countries. But such pressures can be expected in the future. The great need today is for these countries to set aside more park or conservation areas, to provide money for management and protection of the parks and wildlife reserves, and to encourage their citizens to use them.

The participants in this course, and those from 15 other countries taking part in each of the courses held in 1965 and 1966, readily admit their admiration for the United States in pioneering the development of the national parks concept. They are also impressed with the National Park Service's administration of American parks, the planning done for the future, and the quality of interpretive facilities available to the visitor.

Lessons Incorporated

They do not agree that the policies of the park service would necessarily be suited to their particular needs (or that these policies are always best for the United States itself.)

However, the best of the principles learned in visits to the United States, or at these international courses, are being incorporated into the planning of other nations as they develop their parks and conservancy reserves.

In these courses, the National Park Service does not try to hide its own shortcomings. It hopes, in exposing these visitors to the good and the bad, to help them to avoid mistakes in their own programs.

The park and conservation experts from abroad are generally amazed at the amount of public land the United States has set aside for its national park system and the dedication to conservation principles by park rangers and most officials in the National Park Service.

Many countries over the years have asked help from the United States in planning their own national parks or setting up national park and reserves systems. In the past 10 years the United States National Park Service has sent advisers to more than 25 countries. In the last two years, American advisers have been in Turkey, Jordan, Ethiopia, Tanzania, Colombia, Argentina, Peru, Venezuela, Australia, and Thailand. More than 50 countries have recently sent experts to the United States seeking information and guidance in working out park problems.

Private Groups Help

The private sector, through American groups such as the African Wildlife Leadership Foundation, the Conservation Foundation, and the New York Zoological Society, has also assisted a number of countries with national park development and wildlife preservation, chiefly in Africa and Latin America.

The United States also realizes it can learn much from other countries.

The Belgian Congo for many years used its four national parks as laboratories for ecological studies. The present national governments of the two Congoes are maintaining these parks effectively, but with less emphasis on basic scientific research.

England in its nature reserves, Poland and Argentina in their national parks, and Germany with its naturschutzparks also do far more basic scientific research than the United States does in its national parks system.

Nations Cooperate

Most science research in U.S. parks has been oriented to specific problems instead of to basic ecological research.

Several African countries, especially Uganda, Kenya, Zambia, and Tanzania, have developed extensive conservation education programs which allow school children to visit the parks in organized groups.

Countries with common boundaries have in many cases cooperated to establish parks; free sharing of facilities and mutual planning still lies in the future.

Uganda and the Congo have founded national parks on their respective sides of Lake Edward on their common boundary. In the past (but not currently) Zambia and Rhodesia cooperated with parks alongside Victoria Falls. Poland and Czechoslovakia have parks on both sides of the Pieniny River and the Tatra Mountains. Argentina and Brazil have adjacent parks at Iguassú Falls.

In 1932, the United States and Canada decided to set up a Waterton-Glacier International Peace Park on the border shared by the two parks. But for all practical purposes, Canada's Waterton Lakes National Park and the United States' Glacier National Park have been completely separate.

North America's first truly international park was established in 1964. It is on Campobello Island, N.B. There Canada and the United States share the administration and development of Roosevelt Campobello International Park at the site of the summer home of President Franklin D. Roosevelt.

Joint U.S.–Mexican Park

The United States-Mexican border at El Paso is the site of the latest international effort. Mexico has recently completed a pavilion, visitor center, and small park on its side of the Chamizal in Ciudad Juarez. The National Park Service will soon build a half-million dollar visitor center and small park on its side of the border.

The International Union for Conservation of Nature and Natural Resources (IUCN) is a strong advocate of boundary-sharing parks to act as a force for peace.

"It is high time that conservation comes to the aid of politicians in bringing nations together," says Mr. Futehally, who is an IUCN board member.

Plans are under way for a second world conference on national parks, to be held in 1972 at Yellowstone National Park in Wyoming. This will be part of the commemoration of the 100th anniversary of the national park concept, which originated in 1872 with the establishment of Yellowstone as the world's first national park. The first world conference on national parks, held in Seattle in 1962, was attended by 145 delegates from 63 countries.

The keynote speaker, U.S. Secretary of the Interior Stewart L. Udall, called for "a common market of conservation knowledge" and commended the conference for striking "a wholesome note of sanity in a troubled world.

"It is a sign that men are questioning the false gods of materialism and are coming to realize that the natural world lies at the very center of an environment that is both life giving and life promoting." he said. "There is hope in this meeting . . . that the values of the spirit are reasserting their primacy—and this in turn gives fresh hope in other vital areas of human endeavor."

World Trust Proposed

In 1965, at the White House conference on international cooperation, one of the major recommendations was establishment of a world heritage trust to encourage preservation of areas such as Grand Canyon and the Everglades, the Serengeti Plains in Tanzania, Angel Falls in Venezuela, Mt. Everest in Nepal and Tibet, and spectacular animal species.

The proposed trust, the recommendation states, "would be responsible to the world community for the stimulation of international cooperative efforts to identify, establish, develop, and manage the world's superb natural and scenic areas and historic sites for the present and future benefit of the entire world citizenry."

Last year in Amsterdam at the International Congress on Nature and Man, Russell E. Train, president of the Conservation Foundation, urged implementation of the world heritage trust through the activities of the United Nations Educational, Scientific, and Cultural Organization.

Mr. Train said the protection of significant areas is not just a matter of local or even national concern. In his words: "All mankind has a stake in such areas. .. The time has come when this principle must be established at the highest level of international affairs and made the subject of priority action by governments and peoples, individually and collectively.

"The question is no longer whether we can afford to undertake such a program. We cannot afford not to."